CONTENTS

SAFETY

In these *Practical guides*, we have used the internationally accepted signs given below to show when you should pay special attention to safety.

highly flammable

take care! (general warning)

explosive

risk of electric shock

toxic

naked flames prohibited

corrosive

wear eye protection

radioactive

wear hand protection

INTRODUCTION

The practical investigations in this *Guide* relate largely to the topics covered in *Study guide II*, Part Three, 'Inheritance and development', Chapters 15 to 20. Cross references to the *Study guide* are given.

Chapter 15 CELL DEVELOPMENT AND DIFFERENTIATION

Investigation 15A Mitosis in a plant meristem. (*Study guide* 15.2 'Cell division during growth and development'.)
Bulbs and seedlings are used to grow root tips which are stained and squashed to show cells in mitosis.

Investigation 15B Differentiation in stems and roots. (*Study guide* 15.3 'A review of cell diversity'.)
A range of cells in various stages of growth and differentiation is seen in a very small root tip. Sections cut from a stem are used to show the progress of lignification.

Investigation 15C Differentiation in cells. (*Study guide* 15.3 'A review of cell diversity'.)
Highly differentiated, lignified stone cells are observed in close association with parenchyma.

Chapter 16 THE CELL NUCLEUS AND INHERITANCE

Investigation 16A Gamete production by a plant. (*Study guide* 16.4 'Meiosis and its significance'.)
Developing anthers are stained and squashed and pollen mother cells observed.

Investigation 16B Gamete production in an animal. (*Study guide* 16.4 'Meiosis and its significance'.)
A locust is dissected and a squash preparation of the testis used to show development of sperm.

Investigation 16C Inheritance of ability to synthesize starch in *Zea mays*. (*Study guide* 16.6 'Inheritance in fungi'.)
Pollen from a plant which is the offspring of a cross between two strains that produce different types of starch is stained with aqueous iodine, and the grains are scored on the basis of the type of starch they contain.

Investigation 16D Spore colour in *Sordaria fimicola*. (*Study guide* 16.6 'Inheritance in fungi'.)
Asci of *Sordaria* are observed and scored on the basis of the distribution of black and white spores within them. The pattern of spore distribution is related to meiosis.

Chapter 17 VARIATION AND ITS CAUSES

Investigation 17A Describing variation. (*Study guide* 17.1 'Introduction'; Study item 17.11 'Continuous and discontinuous variation'.)
Variation is observed in a suitable organism (*Trifolium repens* and *Homo sapiens* are suggested as examples). Contingency tables, frequency distribution tables, and histograms are used to describe and analyse the variation shown.

Investigation 17B The influence of environment on development. (*Study guide* 17.2 'The role of inheritance and environment'.)
Two situations are observed, one in fruit flies and one in seedlings, in which a change in genotype and a change in the environment can influence the phenotype in similar ways.

Investigation 17C Patterns of inheritance. (*Study guide* 17.3 'Mendel and his contemporaries'; 17.4 'Dihybrid crosses'; 17.5 'Autosomal linkage'; and 17.6 'A model for the inheritance of continuously varying characters'.)
A number of possibilities are suggested for breeding experiments using plants, small mammals, flour beetles and fungi. Techniques for using fruit flies are described in some detail.

Investigation 17D Linkage and linkage mapping in tomato. (*Study guide* 17.5 'Autosomal linkage'.)
A number of commercially available families of tomato seedlings are grown and scored on the basis of three linked pairs of contrasting characters. The data are used to map the loci involved.

Investigation 17E Mutation in yeast. (*Study guide* 17.7 'Mutation'.)
Dried household baker's yeast is cultured and triphenyl tetrazolium chloride is used to distinguish respiratory-deficient mutants.

Investigation 17F The effects of irradiation in plant seeds (*Study guide* 17.7 'Mutation'.)
Tomato seeds carrying the mutant allele of a gene, xanthophyllic-2, which reduces chlorophyll content, are obtained commercially. They have been irradiated with varying doses of gamma radiation. After germination mutation is observed and related to dosage.

Chapter 18 THE NATURE OF GENETIC MATERIAL

Investigation 18A Testing for DNA using the Feulgen technique. (*Study guide* 18.1 'The search for genetic material'.)
Acid hydrolysis and Schiff's reagent are used as a histochemical test for DNA and the results are observed in squash preparations.

Investigation 18B Testing for DNA and RNA using methyl green pyronin. (*Study guide* 18.6 'The breaking of the genetic code'; Study item 18.63. 'The message linking nucleus and cytoplasm'.)
Onion bulb epidermis is stained with methyl green pyronin and the blue–green colour, showing DNA, contrasts with the pink, showing RNA.

Investigation 18C Making a model to illustrate the chemical nature of a gene. (*Study guide* 18.4 'The chemical structure of DNA'; 18.5 'A model of the DNA molecule'; 18.6 'The breaking of the genetic code'; 18.7 'The synthesis of RNA'.)
A model of the DNA molecule is made of card and used to illustrate replication; models of ribonucleotides, a matching-messenger RNA molecule for the DNA model, and other models are also constructed.

Chapter 19 GENE ACTION

Investigation 19A Dwarfism in peas. (*Study guide* 19.3 'Control of transcription in higher organisms'; and see–'Gene activity in polytene chromosomes'; 19.5 'Genes that influence metabolic reactions in Humans–see Study item 19.51 'The influence of thiamine on mutant tomatoes; an analysis of an experiment'.)
A tall and a dwarf strain of pea are grown and some of the seedlings treated with IAA and others with gibberellic acid. Effects on leaf number and on elongation of internodes are observed.

Investigation 19B The biochemistry of cyanogenesis in *Trifolium repens* (white clover). (*Study guide* 19.1 'Mutant complementation in Neurospora'–see Study item 19.11 'Mutations and metabolic pathways'; and 19.6 'Gene expression in heterozygotes'.)
Sodium picrate paper is used to test clover leaves for their ability to release hydrogen cyanide. Plants which prove negative are shown to include some which are a source of an enzyme involved in the release of cyanide, even though they are unable to produce it themselves.

Investigation 19C Gene expression in round and wrinkled peas (*Pisum sativum*.) (*Study guide* 19.1 'Mutant complementation in *Neurospora*'–see Study item 19.11 'Mutations and metabolic pathways'; and 19.9 'Epistasis'.)
The apparently simple phenotypic difference–round or wrinkled–is shown to be related to differences in ability to absorb water, differences in starch grain structure and size, and differences in the level of phosphorylases active in starch synthesis *in vitro.*

Chapter 20 POPULATION GENETICS AND SELECTION

Investigation 20A Models of a gene pool. (*Study guide* 20.2 'The Hardy–Weinberg model'.)
Beads are used to represent alleles in a physical model to illustrate some of the principles which must be grasped to understand selection and gene drift.

Investigation 20B Cyanogenesis and selection in *Trifolium repens* (white clover). (*Study guide* 20.5 'Polymorphism'.)
Strongly cyanogenic and acyanogenic clover plants are used to test the possibility that molluscs may exercise selection in favour of the cyanogenic genotypes when they graze. The influence of frost on the leaves is also investigated.

NOTES ON THESE INVESTIGATIONS

This *Guide* gives instructions for a series of practical investigations in the fields of cell differentiation, variation, and inheritance. You are unlikely to have time to carry out each of these yourself and will sometimes have to rely on evidence collected by others. Many more experimental techniques could be used and many more types of investigation could be done; it is perfectly acceptable for you to carry out other investigations if they suit your particular situation.

Some of the investigations have alternatives, each of which is designed to highlight a similar set of biological principles while using different materials or techniques.

If you have an opportunity to carry out a project as part of your biology course you may be able to adapt one or more of the techniques in this *Guide* to solve a specific problem in the field of genetics.

Several investigations involve the preparation of tissues for observation with a microscope, and it will be assumed that you are familiar with a simple microscope and reasonably practised in using it. Nearly all the stains and other chemicals used in preparing slides are likely to corrode and damage your microscope lenses. You must be very careful to use a minimum amount of stain or mounting medium, not to touch a slide with an objective or condenser lens, and to wipe all microscope parts at the end of each practical.

Cell nuclei and the chromosomes they contain are almost transparent, but they react with a number of different stains, each of which has advantages and disadvantages. Some stains are used as a chemical test for a substance in the cells. Full details of two such staining techniques are given in investigations 18A and 18B. In other cases, the stains are merely used to colour the nuclei and make them visible. This means that the stain to use is a matter of choice; two alternative techniques are given below, and reference will be made to them in other parts of this *Guide*.

Toluidine blue for plant tissues

Procedure

1 Place a small piece of the tissue to be stained on a clean slide. Only use 2 or 3 mm or the cells will be too crowded to be seen properly.

2 Place three drops of 1.0 mol dm^{-3} hydrochloric acid on the tissue.

3 Warm the slide very gently with a small Bunsen flame. Do not let the liquid boil or dry up. After about 1 minute the pieces should appear fuzzy, showing that the cells are separating from each other.

4 Blot off the acid with filter paper or a paper tissue.

5 Add three drops of toluidine blue solution. Warm the slide as before for one minute.

6 Blot off the stain. Add one drop of fresh stain.

7 Mount the specimen as follows.

1 Put the specimen in the centre of the slide.

2 Cover it with a single drop of mountant liquid–propane-1,2,3-triol (glycerol), water, or stain. Use a size of drop appropriate to the size of the specimen and coverslip.

3 Make sure the specimen is wetted and there are no visible air bubbles.

4 Lower the coverslip slowly with needle or forceps as shown in *figure 1.* A thin film of liquid must move over the specimen and leave no air bubbles.

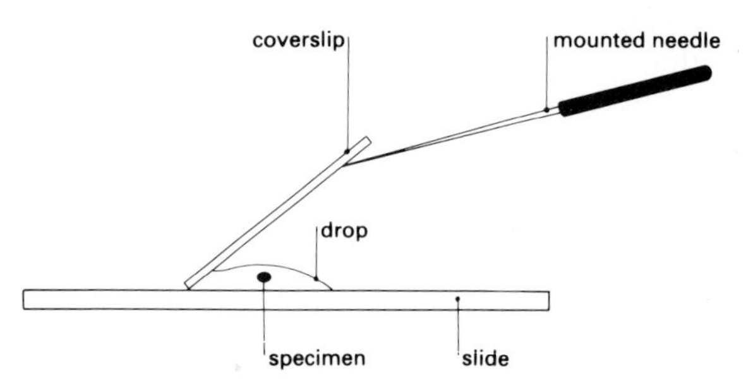

Figure 1
Mounting a specimen.

8 Carry out one of the squash techniques set out opposite.

Ethano-orcein for plant or animal tissues

Fixation coagulates the protoplasm of cells so that it does not disintegrate or change chemically and prepares the tissue to receive the stain. Fixation with ethanoic alcohol is essential before using orcein stain.

Procedure

1 Fix the piece of tissue in a mixture of 1 part glacial ethanoic acid and 3 parts absolute ethanol. The mixture must be fresh. Minimum fixation times for certain tissues you are likely to stain are as follows.
locust testis follicle: 10 minutes
developing anthers: 10 minutes
root tips: 12 hours.

2 Put the fixed tissue into a watch-glass or small beaker and cover with a mixture of 9 parts ethano-orcein stain and 1 part of $1.0\,\text{mol}\,\text{dm}^{-3}$ hydrochloric acid.

3 Warm the mixture and the tissue to 60 °C on a thermostatically controlled hotplate. (If one is not available use a spirit lamp, and keep the watch-glass just warm enough to be painful to touch on the back of the hand.)

4 Keep the temperature between 50 °C and 60 °C for ten minutes. Add fresh drops of stain if evaporation begins to dry out the preparation.

5 Transfer the stained tissue to a clean slide, using a mounted needle. Cut off and discard any parts of the tissue you do not require, and wipe away any stain.

6 Add 1 drop of fresh orcein stain–without any hydrochloric acid.

7 Break up the tissue into very small pieces with a needle.

8 Place a cover slip on the preparation–see *figure 1*.

9 Carry out one of the following squash techniques.

Squash techniques

Procedure

Either

A Place the slide and coverslip on a double layer of filter paper or blotting-paper and fold the paper over the coverslip. Make certain that the slide is on a flat surface, and squash down on the coverslip with strong vertical pressure, using your thumb. Do not twist or roll the thumb from side to side.

Or

B Tap the coverslip about 20 times by dropping a wooden mounted needle or a pencil, blunt end down, from a height of about 5 cm onto the middle of the coverslip. This will disperse the cells and flatten them. You can repeat the process if your observation shows that the tissue is insufficiently squashed.

Whichever squash technique you use, avoid excess stain or pieces of tissue will drift to the edge of the coverslip and be lost. Should many air bubbles appear, add more stain, using a fine dropping pipette, after squashing.

If you have difficulty in squashing you have probably failed to warm your tissue in acid for long enough to separate the cells, or you have too much tissue.

CHAPTER 15 CELL DEVELOPMENT AND DIFFERENTIATION

INVESTIGATION
15A Mitosis in a plant meristem

(*Study guide* 15.2 'Cell division during growth and development'.)

Cell division occurring during the growth of flowering plants takes place in particular regions of the plant called meristems. Cells in meristems are not specialized for any particular function and divide repeatedly by mitosis. Some of the resulting daughter cells remain meristematic, while others cease dividing and become differentiated into an appropriate range of cell types depending on their position.

There is a meristem very near the end of an actively growing root, so root tips provide an easily accessible concentration of cells at different stages of division. Roots from bulbs of onion, garlic, or hyacinth, and from seedlings of peas and beans are all suitable, provided that they are healthy and fast growing. It may be interesting to compare cells from different species.

Procedure

1. Cut off three obviously vigorous root tips from the bulb or seedling provided.
2. Fix in ethanoic alcohol if using ethano-orcein stain.
3. Stain the tips using one of the procedures given on pages 5 and 6.
4. Isolate a tip that has stained well on a clean slide and remove the base (the blunter less stained end) and, if visible, the root cap which may also be less stained.
5. Make a squash preparation using one of the techniques given on page 7.
6. View the slide under the low power (L.P.) of the microscope. Try to locate the meristematic zone, which will be characterized by having small, apparently 'square' cells with nuclei which are large relative to the whole cell area.
7. When you have located the meristematic zone, change to high power (H.P.) and examine the cells. Make representative drawings of any cell that shows any of the stages of mitosis. It is helpful at this stage to refer to published photographs or drawings of mitotic stages, but do not draw from the photographs – use them to help you to identify the stages in your own preparation which you then draw. Annotate your drawings to point out what is occurring in the cell at each stage of the process. Draw several cells fairly quickly. There is

little to be gained from one or two drawings that attempt to show more than your microscope can resolve.

8. If you have squashed the root tip effectively, the cells will be well spread out on the slide. Take a number of fields of view under H.P. and, for each one, count the number of cells at each stage of mitosis. Make a table to show your results and those of other members of the class who have used different root tips.

Figure 2
Nine stages of mitosis in the endosperm of *Tulbaghia* × *rendooznia* (× 1950).
Photographs, Dr. C. G. Vosa, The Botany School, Oxford.

a

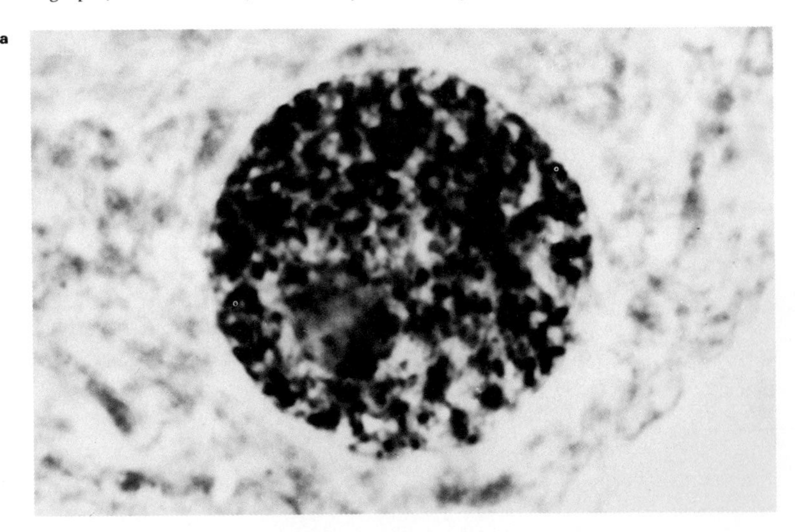

b

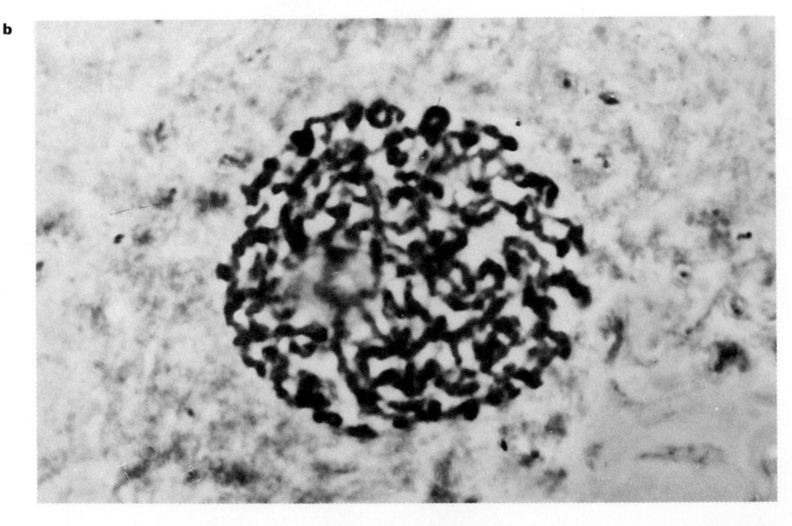

c

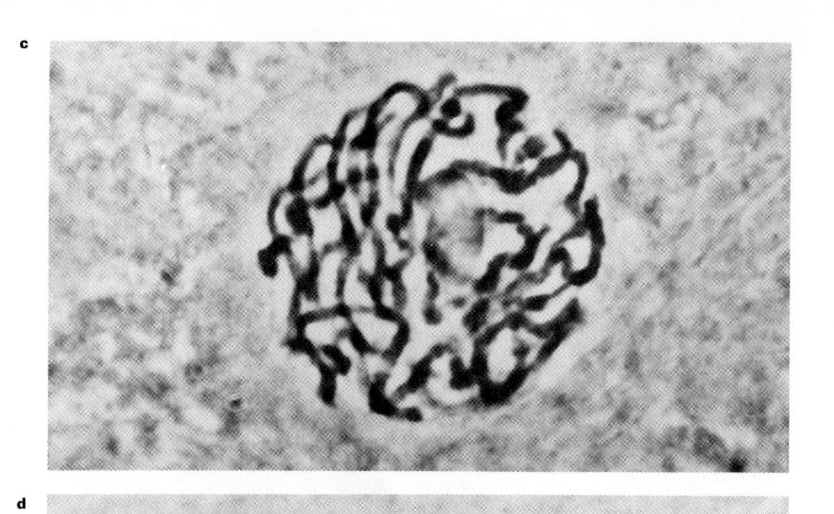

d

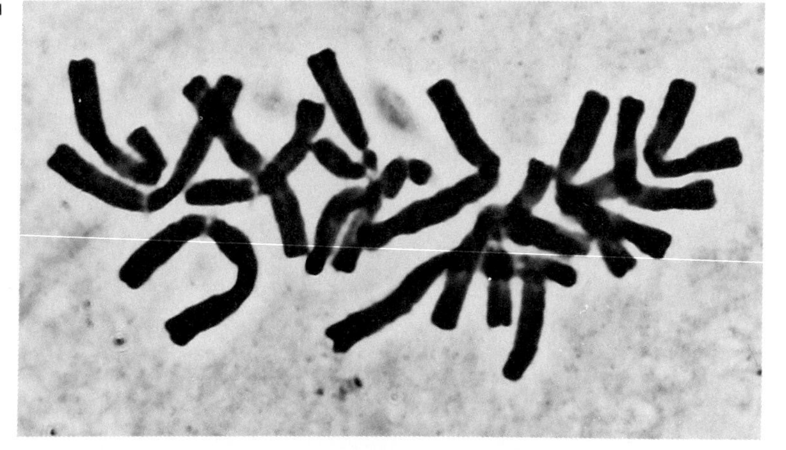

e

f

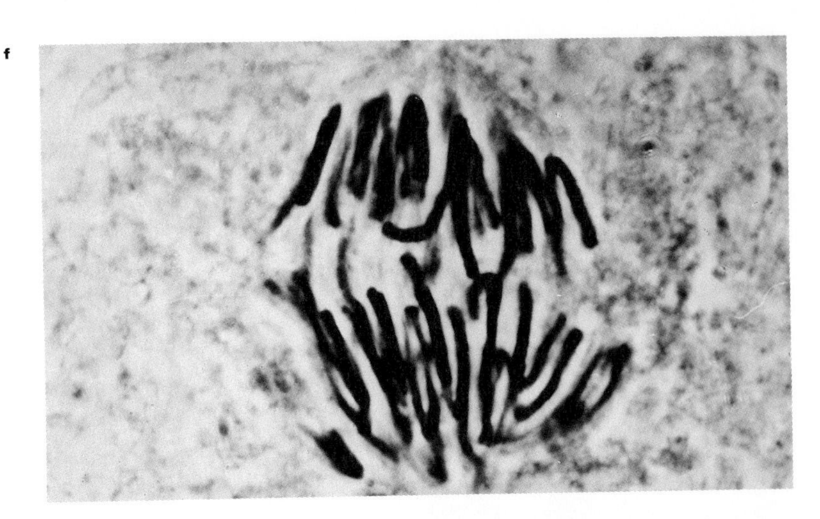

g

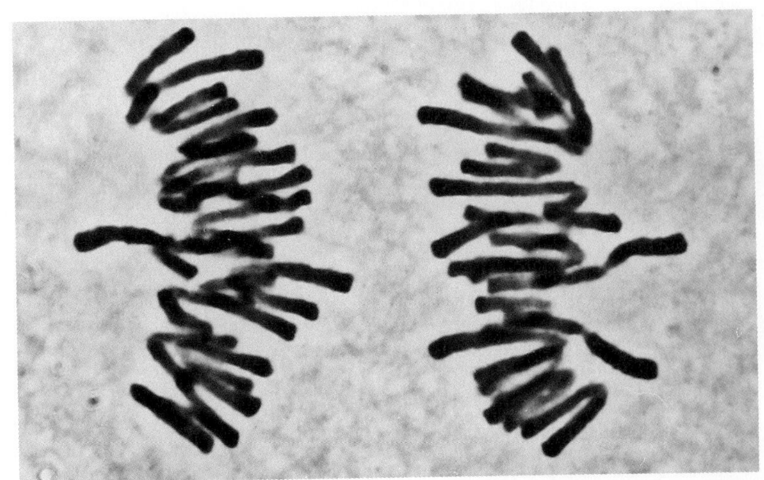

h

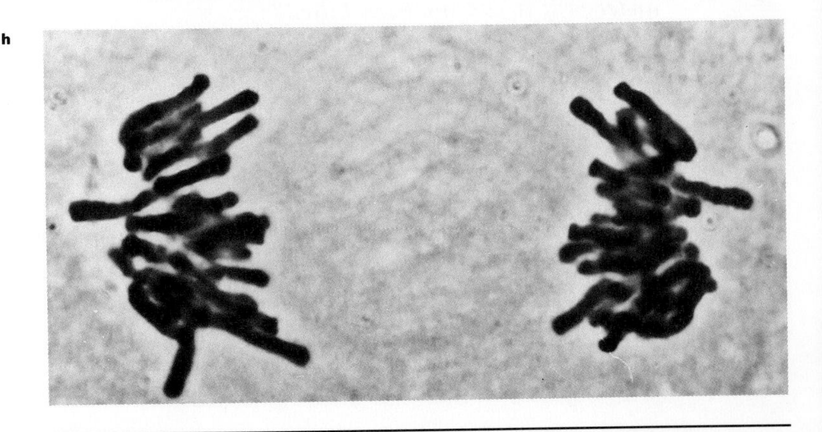

i

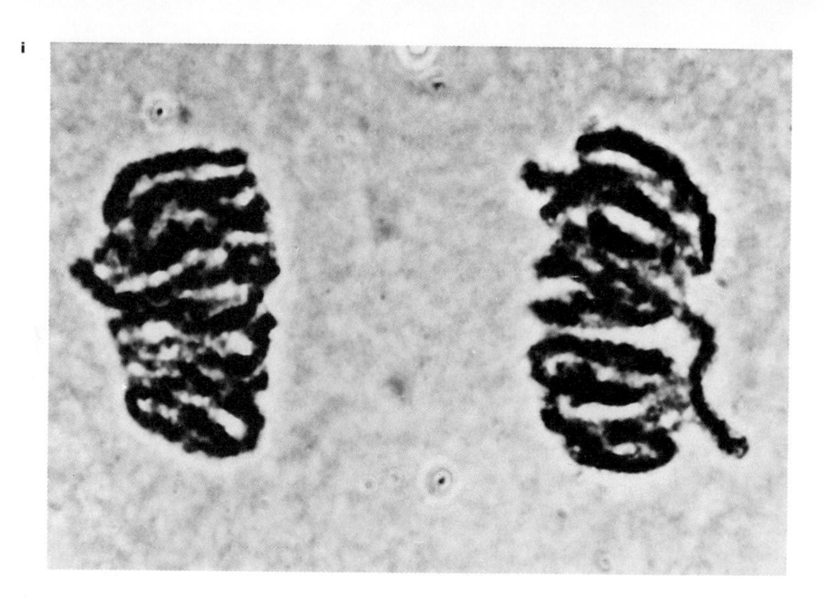

Questions

a **What structure of the cell, essential to the process of mitosis, is not visible in your preparations? Suggest three reasons why this may be so.**

b **Suggest a reason why more cells appear to be in interphase than in any of the active stages of mitosis. What can you infer from your counts of the number of cells seen at each of the four stages of mitosis?**

c **You have been asked to classify cells into different stages of mitosis. Do your slides provide evidence that these stages are real and that they represent a sequence of processes which each cell would undergo if it had not been killed?**

INVESTIGATION 15B Differentiation in stems and roots

(*Study guide* 15.3 'A review of cell diversity'.)

The cells of multicellular plants are contained within polysaccharide walls and cemented together firmly. This means that extensive cell division and growth are not possible within a functioning organ unless its existing tissues are disrupted. Because growth is initiated at the apex, the rest of the organ provides a record of the growth and differentiation of its cells. The further back you go in time, the further you are from the apex.

Many plants develop a secondary meristem called a cambium which results in lateral growth of the stem or root. This investigation will be confined to organs in which little if any cambial activity has taken place.

Procedure A

Some seedlings of tobacco or millet will have been grown on agar in a Petri dish.

1 Remove the lid of the Petri dish and place the dish on a microscope stage, taking care not to touch the agar with an objective lens. View one of the seedlings with a low power objective, and locate the root cap and the root hair region.

2 With a scalpel tip or a needle, transfer the entire seedling to a clean slide, and cut off and discard the seed or expanded cotyledons. Do not damage the root.

3 Stain the root with toluidine blue, following the procedure on page 5. Since the root is so small and thin, use only half the time indicated for acid treatment and staining.

4 Cover the preparation with a coverslip but do not squash. View it using low or medium power.

5 Cells and nuclei will be clearly visible. If necessary, tap the coverslip once or twice to spread the cells slightly but not enough to disrupt their arrangement.

6 Using a graticule, measure the length of 20 cells from the meristem region, and the diameters of their nuclei. Tabulate your results. You will need to devise an unbiased means of selecting the cells.

7 Repeat procedure **6** for the area of root where differentiation, as revealed by a vascular strand and root hairs, has occurred.

8 Repeat procedure **6** for a region between the meristem and the more differentiated region, that is, where cell enlargement may have been in progress.

9 Prepare a table of all the results, and calculate the mean cell length and nuclear diameter for each region measured.

Questions

a ***Did you observe any direct evidence of meristematic activity in the region of very small densely packed cells behind the cap? If not, refer to*** **figure 3.**

b ***Comment on your table of cell lengths and nuclear diameters.***

c ***Have you gathered any evidence for cell wall or cytoplasm changes during the growth of cells?***

Figure 3
a A longitudinal section of a broad bean root tip (× 19).
b Cells from 2 mm behind the root tip (× 145).
c Cells from 1–2 mm behind the root tip (× 145).
d Cells from just behind the root tip (× 145).
Photographs, Philip Harris Biological Ltd.

a

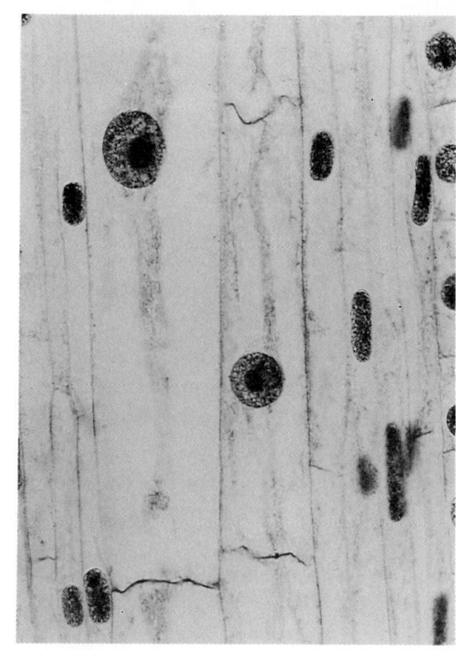

b

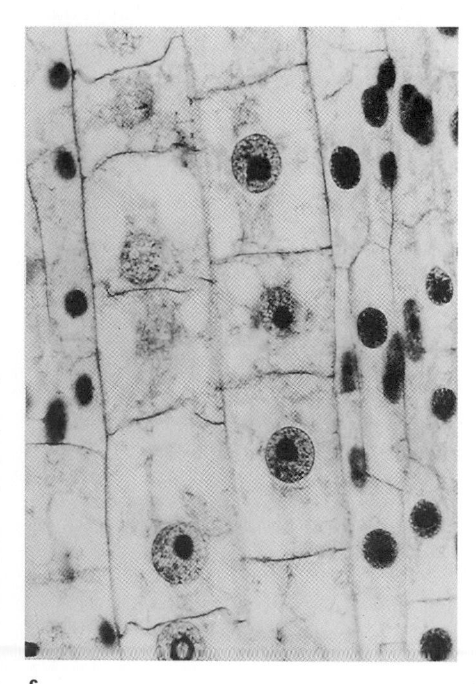

c

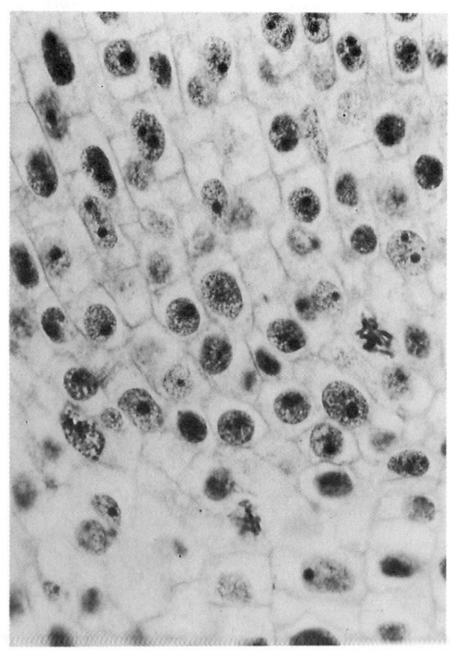

d

Procedure B

Xylem vessels and tracheids are specialized for support and water transport. Fibres, specialized for support only, have thicker walls and a correspondingly smaller lumen. All these cell types have deposits of a complex polymer called lignin in their walls after differentiation has been completed. You will stain for lignin in the procedure below, and use it as an index of differentiation.

1 Choose a vigorous, turgid pea seedling which has developed about 5 to 8 leaves. Cut off the shoot at soil level and transfer to a jar or beaker of water.

2 Count and record the number of internodes from the apical bud to the base.

3 Prepare three Petri dishes or watch-glasses of water, labelled tip, middle, and base.

4 Select an internode near the base of the shoot, record its number, and, using a very sharp (new) razor blade, make a sliding, transverse cut across the internode to give a thin section of tissue (*figure 4a*). The blade should be moistened with water and the section should be as thin as possible. Float the sections in their labelled watch-glass or Petri dish.

5 Repeat procedure **4** for an internode half way up the shoot.

6 Finally, section the very smallest and youngest internode at the apex of the shoot. This will be the most difficult. Do not mistake the petiole of the first leaf for the youngest internode.

7 Transfer a section from the youngest internode to a clean labelled slide. View the section quickly under low power and make sure that the vascular bundles are cut cleanly across and are visible in true transverse section. Ask for advice if in doubt (see also *figure 4b*).

8 Add one drop of phloroglucinol solution to the section and, after 30 seconds, one small drop of concentrated hydrochloric acid.

9 Add a coverslip and view under low power.

10 Lignified cell walls will be red, and the cells are likely to be smaller in diameter than most of the surrounding unlignified ones. If cells are pale pink, then slight lignification has occurred. If pink or red strands or spring-like structures are visible, a section of vascular bundle has collapsed and is being viewed longitudinally and not transversely.

11 Count the number of vascular bundles, and estimate as accurately as possible the number of lignified cells in the section.

12 Count the number of cells in one diameter of the section, using the eyepiece graticule as a guide, and measure the diameter.

13 Repeat procedures **7** to **12** for the oldest internode section you have cut, and then for the one of intermediate age.

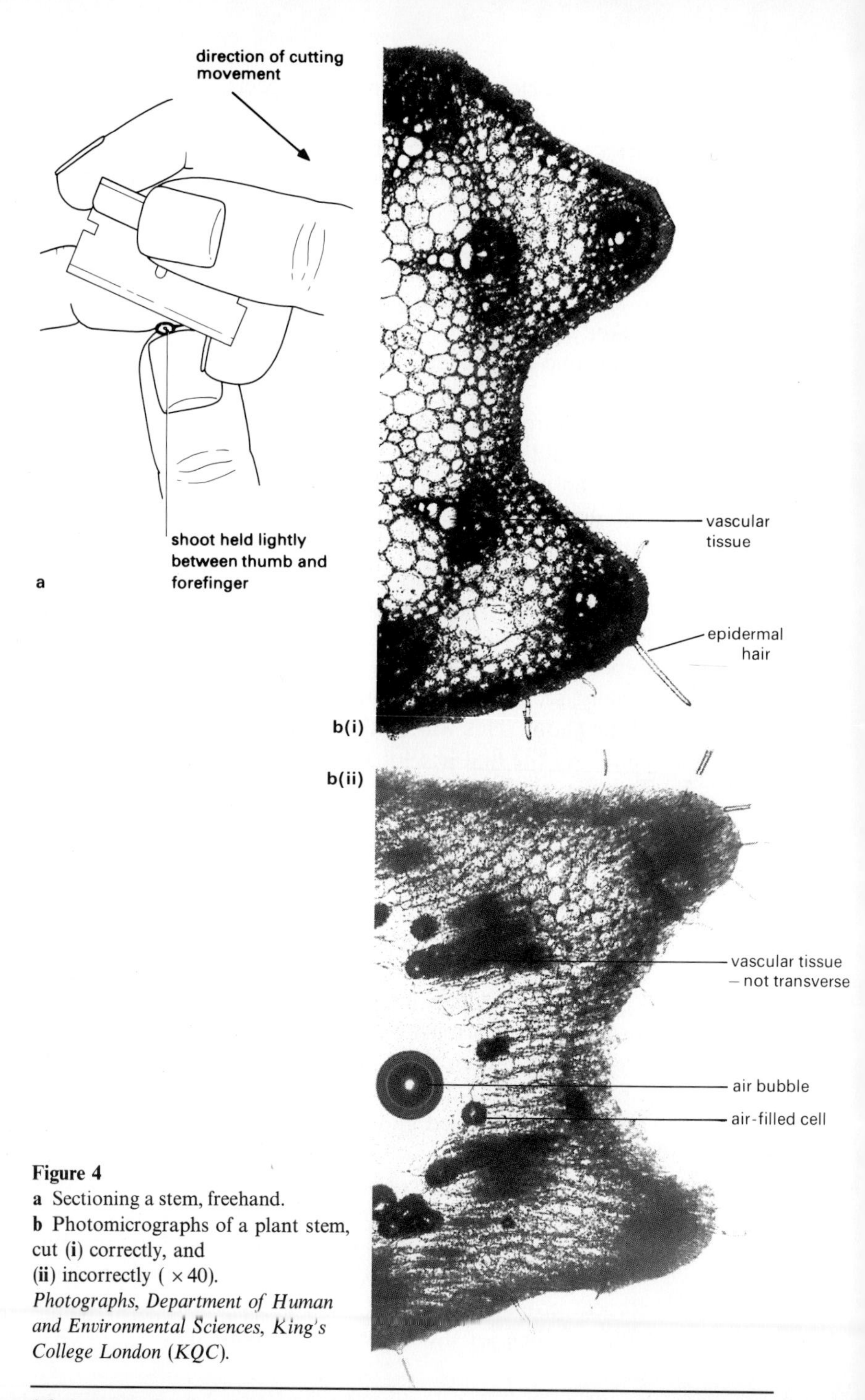

Figure 4
a Sectioning a stem, freehand.
b Photomicrographs of a plant stem, cut (**i**) correctly, and
(**ii**) incorrectly (×40).
Photographs, Department of Human and Environmental Sciences, King's College London (KQC).

Questions

a **_Use your data to argue whether growth between the youngest and the oldest internodes which you observed has been largely by cell division or largely by cell enlargement._**

b **_What evidence do you have that lignification of cells increases progressively with the age of the stem?_**

INVESTIGATION
15C Differentiation in cells

(*Study guide* 15.3 'A review of cell diversity'.)

Plant cells are derived by mitosis from cells in meristematic regions of the plant. Before a newly divided plant cell can perform specialized functions it must first enlarge (by a process of vacuolation) and differentiate. Differentiation involves both structural and functional changes in the cell. During differentiation, modifications to the cell wall frequently occur; these may simply involve the addition of further layers of cellulose to the primary cellulose wall, or there may be the addition or superposition of lignin in various patterns onto the primary wall.

Cellulose is a simple polysaccharide; lignin is a complex mixture of sugars, aromatic molecules, amino acids, and so on. Simple tests can be performed to determine the distribution of these two molecules in plant tissue.

Procedure A: plant cells

1. Cut a few sections of the 'flesh' of a pear fruit. The sections should be as thin as possible. Leave them in water in a watch-glass.
2. Label two slides A and B. Onto slide A put 4 drops of iodine in potassium iodide solution and immerse a very small section of the pear tissue in it. Transfer the tissue by lifting it on a mounted needle or section lifter.
3. Allow the tissue to soak for 1 minute before taking up the excess solution (about half of its volume) with absorbent paper.

4. Very carefully add two drops of concentrated sulphuric acid directly onto the section and rock the slide back and forth to ensure good mixing.
5. This technique will stain cellulose blue. It may be that not all the cellulose will stain blue in any one part of the section, depending upon its thickness, but cells at the edge of the section will quickly become deeply coloured. Place a coverslip on the slide and examine under low or medium power of a microscope.

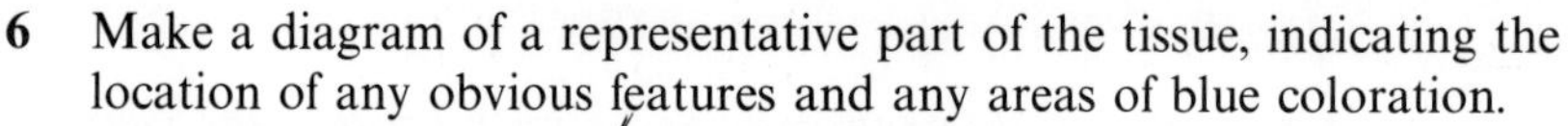

6 Make a diagram of a representative part of the tissue, indicating the location of any obvious features and any areas of blue coloration.

7 Onto slide B put 4 drops of phloroglucinol and place another section of the pear material in this. Rock the slide to ensure that the section is thoroughly wetted by the chemical.

8 Inspect the tissue with the naked eye. If you cannot see any obvious small red patches (lignin stains red), add a drop of concentrated hydrochloric acid to the section to encourage staining.

9 Place a coverslip on the slide and examine under low or medium power of a microscope.

10 Make a diagram of a representative part of the tissue, showing the areas of red coloration.

11 From slide B draw, in detail from H.P., one of the red-stained cells (which are called sclereids) and one of the unstained cells (parenchyma) that you can see.

Questions

a ***It is easier to explain the occurrence of a homogeneous differentiated tissue than it is to explain how the scattered patches of sclerenchyma (sclereids) as seen here may have arisen. What are the problems that have to be solved in order to explain the distribution of the sclereids?***

b ***The lines on the cell wall that you observed in step 11 above are called 'pits'. Suggest a function that they might perform and explain why they are tubular.***

Procedure B: animal cells

1 Examine electronmicrographs of two animal cells. They will be from different tissues and will have differentiated in different ways.

2 Describe the features that you can observe in each of the two cells that are absent from the other cell, and suggest or find out a function of the features. It will help if you bear continually in mind the functions of the tissue or organs from which the cells came.

Questions

c ***Explain the difference between the type of differentiation observed in plant cells such as those seen in part A and in these animal cells.***

d ***What does differentiation achieve 1 for a cell; and 2 for an organism?***

CHAPTER 16 **THE CELL NUCLEUS AND INHERITANCE**

INVESTIGATION
16A Gamete production by a plant

(*Study guide* 16.4 'Meiosis and its significance'.)

Pollen is manufactured in the anthers within the flowers of most angiosperms. Even while the flower is still in the form of a very young bud, the anthers are forming, and by the time that a clearly recognizable flower bud or flower has grown, the pollen grains have been produced and are present in the locules of the anther. You need to examine cells that are currently manufacturing pollen, so it is important to obtain very young anthers for study, in which the pollen mother cells, on the inner walls of the anther, are still active. Fortunately anthers are easy to recognize (see *figure 5* and the *cover photograph*) and so it is possible to locate and remove anthers in very young buds within bulbs, or within recently formed inflorescences (groups of flower buds) on plants coming into flower. These anthers can then be stained to show the structures of the active cells within them.

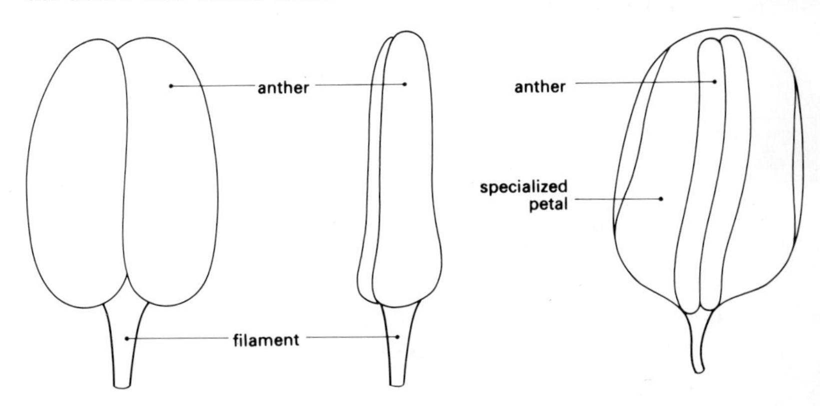

Figure 5
Typical anthers (× 20; × 15; × 10).

Procedure

Location of anthers in a bulb

1 Cut vertically down through the bulb just to one side of the apex.
2 Discard the portion of the bulb that does not contain the bud; the bud is usually yellow–green and is in the centre of the bulb.

3 Remove the white, fleshy leaf bases from around the bud and discard them.
4 With a mounted needle, 'unwrap' the flower bud by peeling away the yellow leaves which tightly enclose it. Now proceed to pick away the petals of the flower until you find the anthers which may be attached to tiny petals or standing free in the centre of the flower, around the style.
5 Proceed to 'Preparation of anthers' below.

Location of anthers in an inflorescence

(It may be helpful to perform the next three steps under a stereo microscope.)

1 Select the very smallest single bud on the inflorescence, which will also be the youngest.
2 Carefully hold the base or stalk of the bud with forceps and pick the green sepals off with a mounted needle.
3 Remove the small white petals that are around the outside of the flower and expose the ring of anthers in the middle of the flower. In some flowers, the anthers may themselves be attached to very small petals; do not discard these by mistake.

Preparation of anthers

If the anthers are yellow and obviously pollen-filled, they will be too mature for cell division to be taking place. They should appear white and translucent. If in doubt, try a smaller bud. Leave them attached to the flower throughout the staining process as they are very easily lost.

1 Fix in ethanoic alcohol if using ethano-orcein stain.
2 Stain, using one of the techniques described on pages 5 and 6.
3 Detach two or three anthers and transfer to a clean slide.
4 Break up with a mounted needle and discard the filament and any other relatively unstained portions.
5 Make a squash preparation (see page 7).
6 Search the slide and try first to get a general idea of the apparently different types of cell that you can see. There may well be mature or maturing pollen grains and cells showing almost homogeneous nuclei, as well as cells clearly undergoing division. In the cells that are actively dividing, the nuclei may show obvious thread-like structures (chromosomes). Damage to the nucleus or incomplete staining may have occurred. Compare several cells which seem to be at different stages of cell division.
7 Carefully draw three or four different cells, each showing some nuclear structure, but at different stages. Focus carefully up and down with the fine focus as you draw; try to see the whole length of the chromosomes.

Questions

a ***You will almost certainly see some maturing pollen grains on your slide. Compare these with the cells that are clearly in the process of division. What are the major differences between the two types of cell?***

b ***State any evidence which you saw in your preparation which led you to the conclusion that the cells were undergoing an orderly process rather than a random division of contents. (Cell division in gamete-producing cells is called meiosis.)***

c ***Compare the actively dividing cells which you see in this preparation with those which you observed in a root tip preparation. Can you see anything which would lead you to believe that a different process was occurring in the two types of cell?***

INTRODUCTION
16B Gamete production in an animal

(*Study guide* 16.4 'Meiosis and its significance'.)

The sperm of a locust are manufactured in the testis which is a fat-covered organ lying on top of the gut in the animal's abdomen. The testis consists of many individual finger-like lobes called follicles, each 1 to 2 mm in length and 0.1 mm in width, in which the cell division necessary for gamete production occurs.

Production of sperm starts early in the animal's developmental life and is completed very early in the life of the adult. There is a certain degree of synchrony about the gamete production in any one follicle, whose cells tend to be at roughly the same stage in their development.

Procedure

1 Take a freshly killed male locust nymph at 3–5 instar stage (see *figure* 6).
2 Cut horizontally across the tip of the abdomen so that the point of a pair of scissors may be inserted.
3 Hold the locust in one hand and carefully cut along one side of the abdominal wall, taking care to keep the scissors pointing outwards so that you do not cut any internal organs. Cut up as far as the thorax.
4 Put the locust into a wax dish and pin it through the thorax. Open the abdominal cavity and pin the integument out dorsally and ventrally (see *figure* 7). Flood the preparation with insect saline.
5 Look for and locate the testis. You should find it lying on top of the dark brown gut in the 4th to 6th abdominal segment (from the

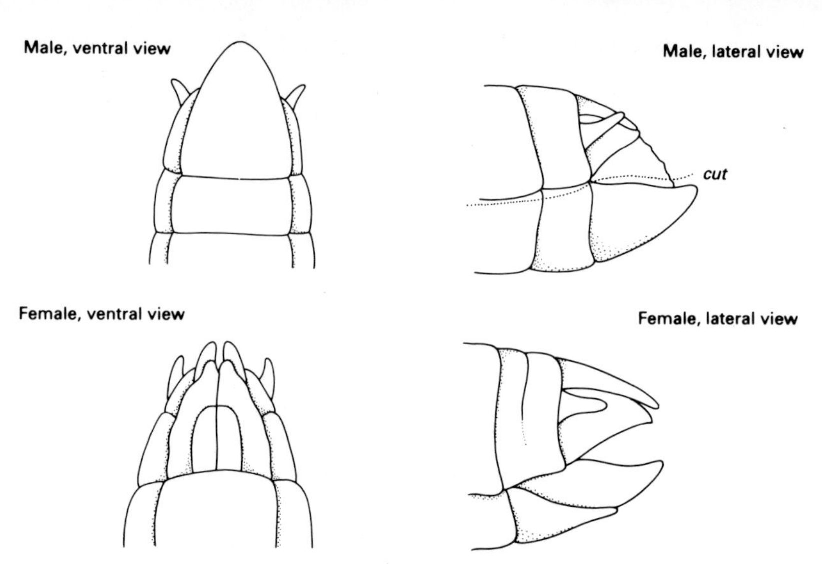

Figure 6
The identification of the male locust. The diagnostic feature is the underside of the terminal segment of the abdomen. In the male this is markedly convex and is undivided. In the female it is less markedly convex and there is a division along the midline. (All drawings × 5.)

Figure 7
A dissection of a locust showing the testis in position on the dorsal surface of the gut (× 2.5).

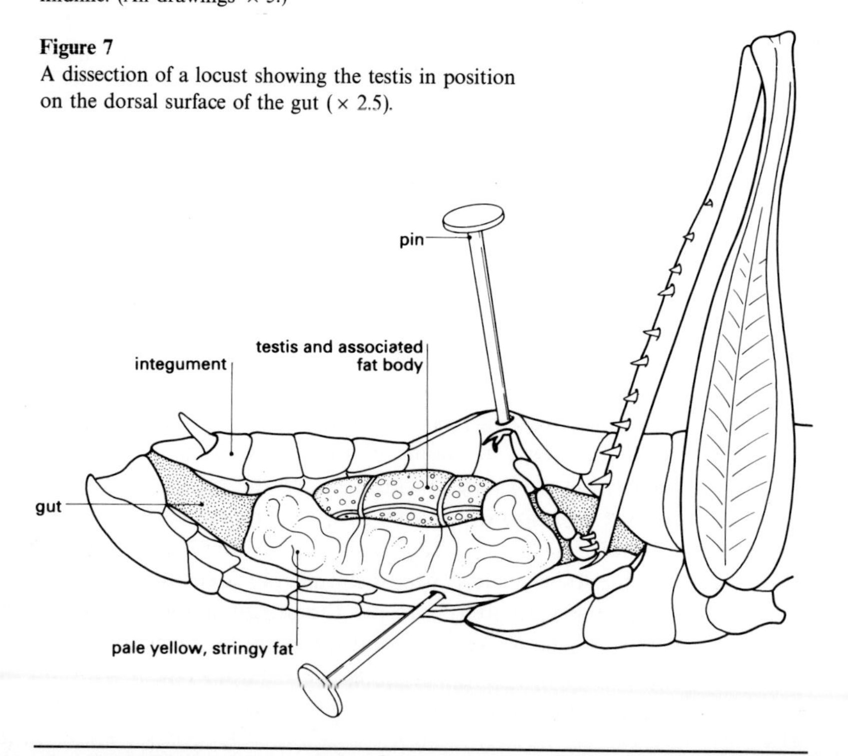

anterior end). There will perhaps be quite a lot of very pale yellow, stringy fat around it, but the testis and its fat constitute a bright yellow structure whose outline is clearly recognizable (*figure 8*).

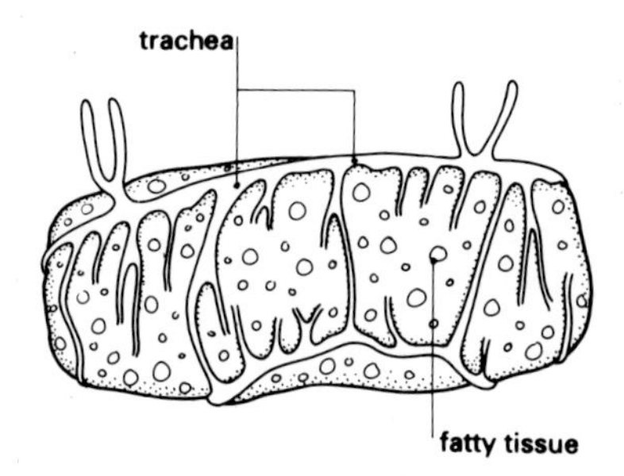

Figure 8
The locust's testis and associated fat body, showing the typical shape (× 6).

6 Carefully pull this yellow structure away from the gut with forceps and remove any of the fat 'strings' which may cling to it.
7 With two pairs of forceps pull the testis and its fat body into small pieces and carefully separate the bright yellow fat from the very pale white, finger-like follicles that are revealed as you pull the structure apart (*figure 9*).

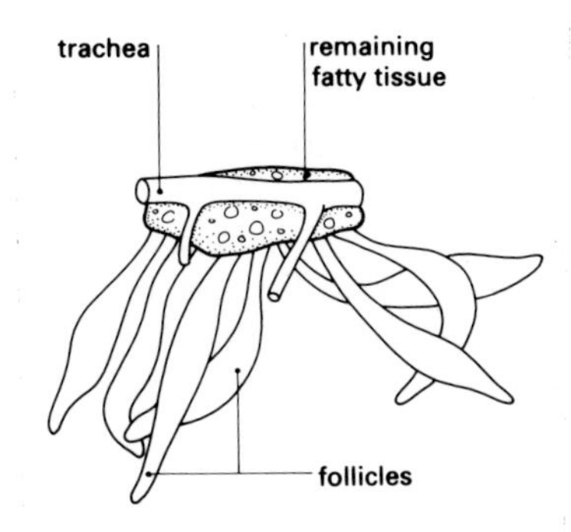

Figure 9
A small piece of the locust's testis after teasing out, showing the follicles (× 23).

8 Fix the follicles in ethanoic alcohol (see page 6).
9 Stain them with ethano-orcein (see page 6).
10 Put three or four stained follicles on a clean slide, break them up with a mounted needle, and make a squash preparation (see page 7).
11 Search the slide thoroughly and attempt to identify chromosomes within nuclei. Remember that the nuclei are unlikely to be completely flat, so chromosomes may not lie with their complete length in any one focal plane.

12 Make representative drawings of four or five of the different types of nucleus or stages of division that are visible under high power and label each drawing fully. The cells in any one follicle are likely to be all at the same stage, so it is useful to examine several follicles.

Questions

a ***What evidence did you see that indicated that these cells were undergoing an ordered process of division rather than division of a random nature?***

b ***Suggest reasons why it is of advantage to the animal for its testis to mature during its nymphal life rather than to begin its maturing process only after the animal is adult.***

c ***If it is possible, arrange the nuclei that you observed into some sort of order representing progressive change. What is the significance of the order into which the nuclei fit?***

INVESTIGATION
16C Inheritance of ability to synthesize starch in *Zea mays*

(*Study guide* 16.6 'Inheritance in fungi'.)

Plants of *Zea mays* (maize) bear separate male and female flowers on the same plant and the flowers occur in groups known as inflorescences. The male inflorescence is called a spike and the flowers are in pairs called spikelets. There are no petals, and the male sexual parts are protected by leaf-like outer sheaths and papery thin inner sheaths. Within this protection will be found the large anthers which, in the mature flower, hang outside the sheaths for pollen dispersal by the wind. The female inflorescence has no individual protective parts around each flower or pair of flowers, and this means that the mature inflorescence, after fertilization, consists of rows of unprotected fruits – the familiar sweet-corn cob.

Starch is a mixture of two types of chain made up of linked glucose units. Amylose is an unbranched chain which stains blue–black with aqueous iodine solution. Amylopectin is a much branched chain, staining red–brown with iodine solution. In starch from different plants there are various proportions of these two polymers (each type also varying in the length of chain).

True breeding strains of maize are available which differ in the type of starch they produce. 'Starchy' strains produce amylose and amylopectin and their tissues stain blue (the blue–black colour masking the red–brown). 'Waxy' strains produce only amylopectin (with no amylose) and so stain reddish–brown with iodine solution.

If a starchy and a waxy strain are crossed, the hybrid offspring are all starchy in character. This investigation examines the polysaccharide reserves of the pollen produced by hybrid plants.

Procedure

You are provided with a portion of male inflorescence taken from a hybrid plant. The plant originated from a cross between true-breeding waxy and starchy varieties.

1 Remove a spikelet from the inflorescence with forceps and place it on a slide. (The spikelet should be closed, but preferably not too near the top of the inflorescence as it may not then have matured sufficiently.)

2 Hold the spikelet by the base with forceps. With a second pair of forceps remove the opaque sheaths that enclose the flowers.

3 The anthers are visible as bright yellow, tubular structures within the flowers, and there are three such anthers in each of the two flowers which make up the spikelet. Crush about one half of an anther onto the centre of a dry slide with a mounted needle.

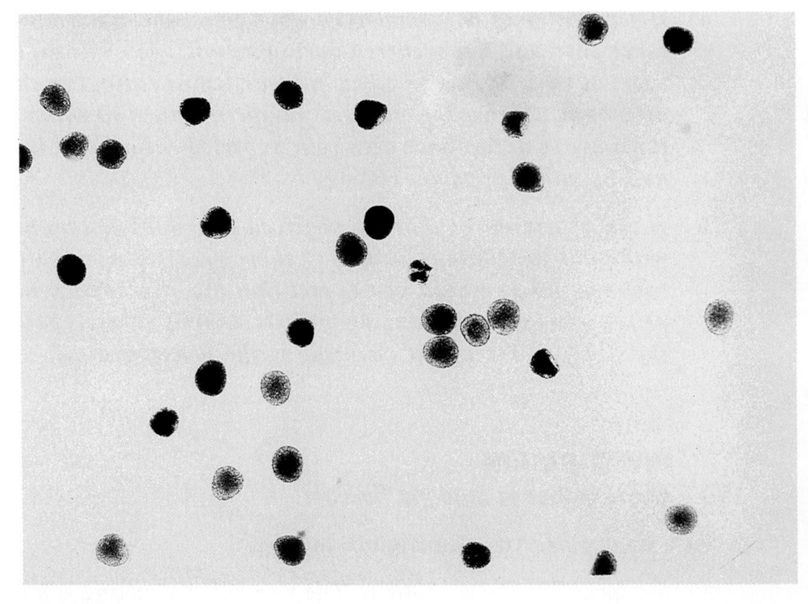

Figure 10
Photomicrograph of iodine test applied to pollen from maize heterozygous for the waxy gene (× 70).
Photograph, Department of Human and Environmental Sciences, King's College London (KQC).

4 Place one drop of iodine solution onto the crushed anther.
5 Stir the pieces of anther and the pollen in the iodine solution with a mounted needle, remove the large pieces of anther, and leave for 5–10 minutes to take up the stain.
6 Place a coverslip on the preparation and view under low power.
7 Adjust the mirror so that light is not passing directly up the microscope tube and so the pollen grains are illuminated from the side of the field of view. If the grains all look the same, you may have to try again with a different concentration of iodine solution, and change the angle of illumination, or the light intensity, or the magnification of the objective lens. With a suitable combination of lighting and iodine concentration you will see two types of pollen. Those only containing amylopectin will appear brownish. Those with any amylose look dark grey to blue–black in colour. (See *figure 10*.)
8 Count the numbers of the two types of grain in many different fields of view, taking care not to count the same grain twice. (Some grains may be shrivelled in appearance or much smaller than the others – do not score these abnormal grains.)

Questions

a ***If separation of homologous chromosomes, one maternal and the other paternal, has occurred during meiosis, you will expect the two types of pollen grain to occur in a particular ratio. Use the χ^2 test to determine whether the observed numbers of the two pollen grain types are in accordance with your expected ratio. State in words the null hypothesis you are testing.***

b ***If the χ^2 test shows that the observed distribution does not agree with your predictions, assuming those predictions to be genetically 'correct', where would you expect that the discrepancy had first arisen – in the anther during meiosis, during your preparation of the slide, or during the counting of the pollen grains?***

INVESTIGATION
16D Spore colour in *Sordaria fimicola*

(*Study guide* 16.6 'Inheritance in fungi'.)

Sordaria fimicola is a fungus. The vegetative structure is a mycelium of multinucleate, haploid hyphae. A mycelium grown from a single spore contains nuclei all of which are genetically identical as they are produced by mitosis of the original nucleus giving rise to the spore. Such a mycelium is called a *homokaryon.*

As they grow through a nutrient medium, hyphae from different mycelia may meet and fuse to give a haploid mycelium containing two genetically different types of nuclei. This is called a *heterokaryon.*

Under appropriate conditions, hyphae differentiate into flask-shaped structures each called a *perithecium* (plural perithecia). The perithecia are about the size of pinheads and become very darkly pigmented as they mature. Inside the developing perithecium, specialized binucleate cells form and within these, nuclear fusion occurs. This results in a number of diploid cells. Each diploid cell elongates to produce a long thin cell called an ascus. Meiosis takes place within the ascus and each of the four haploid products of meiosis divide once by mitosis. This results in pairs of genetically identical nuclei. Because the ascus is a thin cylindrical 'sleeve' the nuclei must remain in a line, unable to slide past each other. As nuclear divisions proceed, the two products of each division are always adjacent to each other. A diagram of the life cycle of *Sordaria*, highly simplified, is shown in *figure 11.*

Figure 11
The life cycle of *Sordaria fimicola.*
(The diagram is not to scale.)

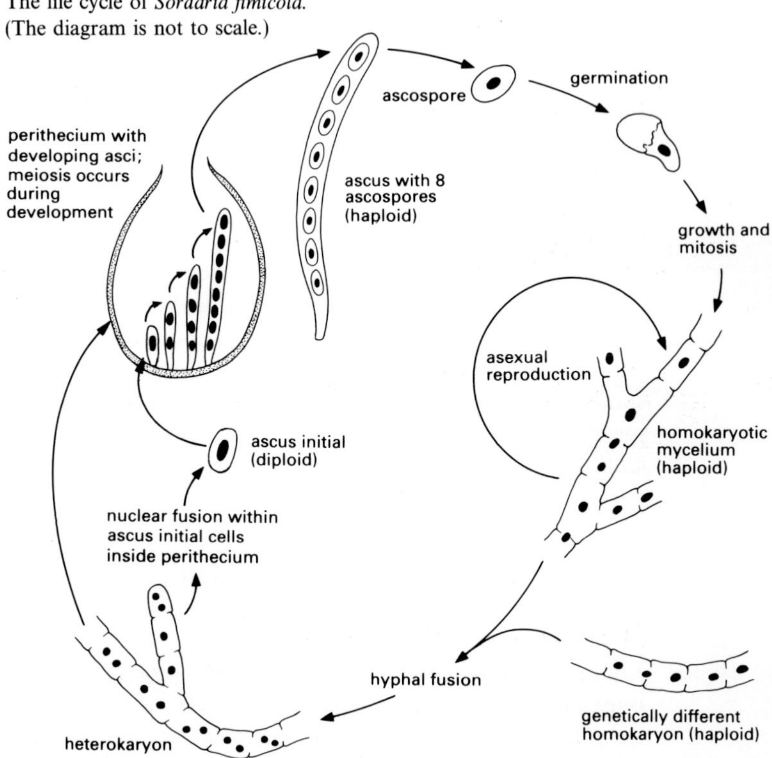

The nuclear divisions that occur as an ascus develops are shown schematically in *figure 12.*

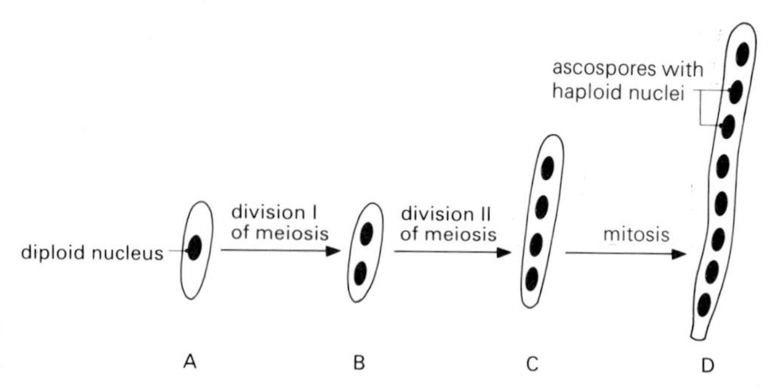

Figure 12
Stages in the development of an ascus.

Each haploid nucleus in an ascus results in a haploid spore so that the mature ascus contains eight ascospores. These are ultimately shot some distance through an opening at the top of the perithecium, and so dispersed.

Perithecia may be squashed and complete asci recovered before their spores are discharged. Each spore in turn may be removed from a row of eight by micro-dissection and transferred to a separate Petri dish of medium where it will develop into a homokaryotic mycelium.

The genetic constitution of each spore in the ascus can thus be determined and hence the genetic consequences of a single meiosis can be observed.

The ascospores of wild type *Sordaria* are black. A variant of the fungus produces white spores. The difference between black spores and white depends on a single pair of alleles, b^+ = black and b = white. If a black spored strain is crossed with a white spored strain, the number and position of black and white spores in asci resulting from the cross can be seen if a hybrid perithecium is squashed.

Procedure

(Steps **1** to **4** may be done for you.)

1 Mark the underside of a sterile Petri dish containing cornmeal agar with two crosses on a diameter of the dish, lying about 2.5 cm apart. Mark one cross ‘bl’ and the other ‘w’.

2 Flame a mounted needle and take a small piece of mycelium from a pure culture of unpigmented (white) spore strain of the fungus and transfer it to the spot marked ‘w’ on the sterile dish. Reflame the needle.

3 Repeat step **2**, but this time use the pure culture of the pigmented (black) spore strain and inoculate the spot marked 'bl'. Flame the needle again.

4 Seal the inoculated dish with tape, invert it, and incubate it at 25 °C for 10 days.
The perithecia of both black and white spored strains are black. They tend to form most frequently along a line where the two haploid mycelia meet as they grow outwards from the points of inoculation. Self fertilization may sometimes occur and this results in a perithecium where all spores are of one colour. Immature black spores are colourless and may at first be mistaken for white spores, but mature white spores have a distinctive 'pearly' appearance.

5 With a flamed mounted needle, lift one or two perithecia from the agar and place them in a drop of water on a microscope slide.

6 Add a coverslip to the preparation and tap it gently with the handle of the mounted needle to rupture the perithecia and expose the rosettes of asci.

7 Examine the asci under low power. If the ascospores are immature or are all of one colour, repeat procedure **5** and try again.

8 Record in a table the sequence of black and white spores in as many mature and intact hybrid asci as you find. In each case, call the spore at the base of the ascus (nearest the centre of the rosette of asci) number 1, and the one at the tip of the ascus number 8.

Questions

a ***Make a series of simple diagrams to show the different arrangements of ascospores that are apparent within the asci. Number your diagrams 1, 2, etc.***

b ***Which of your diagrams can be explained by the separation of b and b^+ at the first anaphase of meiosis?***

c ***Which of your diagrams can be explained by the separation of b and b^+ at the second anaphase of meiosis?***

Examine data collected from a whole class.

d ***How do the class data provide evidence that, during meiosis, the orientation of chromosomes on the spindle is random with respect to the two poles of the spindle? (See* Study guide II, *figure 35.)***

At the start of first prophase, the chromosome from the black-spored strain, carrying the allele b^+, had duplicated to produce two chromatids attached by a common centromere. The same was true of the chromosome from the white-spored parent. The centromeres do not divide until the second anaphase.

e ***How can the separation of b and b^+ at the first anaphase of meiosis in some asci, and at the second anaphase of meiosis in other asci, be explained?***

f ***In what percentage of asci did separation occur at the second anaphase of meiosis?***

CHAPTER 17 VARIATION AND ITS CAUSES

INVESTIGATION
17A Describing variation

(*Study guide* 17.1 'Introduction'; Study item 17.11 'Continuous and discontinuous variation'.)

It is probable that no two organisms are exactly the same; however some groups of individuals have characters in common which mark them out as distinct from other groups. The importance of variation is discussed in *Study guide II* page 60.

Variation may be described as continuous or discontinuous. Continuous variation is a matter of degree; individuals cannot be separated into clear cut groups and the difference between them must be described by some sort of measurement. If two exactly similar measurements are found it is probable that a more accurate technique would reveal a difference. Discontinuous variation occurs where observers can reliably assign individuals to different groups, depending on whether or not they show a particular character.

Two alternative investigations are suggested below but almost any type of organism provides suitable material and you may decide to follow the general principles of these investigations but to use a different animal or plant.

Part A Variation in clover

Before you start the exercise, remind yourself of or find out the meaning of the terms node, internode, petiole, leaf, leaflet, stolon, and terminal bud.

You will require polythene bags for collecting clover, and a ruler.

Procedure

1 Select an area containing many patches and clumps of *Trifolium repens.* (Recently mown turf will not be suitable.)
2 From each patch obtain a few stolons with a terminal bud and some mature leaves. Carefully disentangle the petioles from the surrounding vegetation before breaking the stolons.
3 Choose two patches that appear to differ in leaf size and collect a larger number of stolons from these patches than from the others in your sample.
4 Put samples from each patch of clover in a separate polythene bag and seal this with a wire bag fastener to prevent wilting.

5 On returning to the laboratory, place each sample in a separate beaker of water on the bench.

6 Examine the samples collected by the whole class and make a list of characters which vary. Decide in each case whether the variations are continuous or discontinuous.

7 Select one character which shows clear discontinuous variation and score all the samples available for this character.

8 Prepare a contingency table to record your results.

9 Examine the two larger samples of clover which appeared to differ in leaf size.

10 Choose class intervals so that all leaflets subsequently measured from these clumps may be assigned to one of about eight classes of size.

11 For larger samples measure the central leaflet of the first three fully expanded leaves on each stolon. Record your data by a tally system as illustrated below, using one tally for each clump of clover.

Class interval (*mm*) < 3.5 |||
3.6–5.5 |||
5.6–7.5 ~~||||~~ ||||
7.6–9.5 ~~||||~~ |
9.6–11.5 ~~||||~~
11.6–13.5 |||
>13.6 |

12 Convert your tallies to frequency distribution tables and histograms.

13 Calculate the mean and standard deviation for each leaf sample and the standard error of difference of the two samples.

Questions

a ***In these investigations a patch or clump of clover is considered to be one individual. Is this assumption justified?***

b ***Consider the list you have prepared of the ways in which*** **T. repens** ***varies and produce hypotheses to explain why some of the contrasting types you have encountered have survived in the habitats from which the plants were obtained.***

c ***Does your contingency table describing the discontinuous variation in the clover population for one character agree with the tables produced by others in the class who scored the same character? If there is substantial disagreement, try and explain this.***

d ***Why, in each of a sample of stolons, were the first three leaves measured rather than a random sample gathered from each clump?***

e ***Did measurement of length of leaflet in the two clumps of clover reveal a significant difference? How could it be established whether such a difference is caused by environmental factors or by genetic differences?***

Part B Variation in humans

Procedure

1 Select a group of human subjects for study. The group should have a minimum of thirty individuals.
2 Make a list of characters which vary within your chosen group and which you can yourself observe or measure. Decide in each case whether the variation is continuous or discontinuous.
3 Select one character which shows clear discontinuous variation and score all the subjects for this character.
4 Prepare a contingency table to record your results.
5 Ask each subject to stare fixedly ahead and measure the distance between the pupils of his or her eyes in millimetres to the nearest millimetre. Take care that he or she is focused on a distant object and is not squinting at the ruler.
6 On your table of data record the age of each subject to the nearest year and the sex.
7 Inspect your data and choose class intervals so that each subject can be allocated to one of about eight classes of pupil-to-pupil distance.
8 Prepare a tally of your data (see step **11** of part A). This tally may be used to include additional subjects in the survey if time permits. If your sample is large enough you should make a separate tally of males and females.
9 Convert your tally or tallies to frequency distribution tables and histograms.
10 Calculate the mean and standard deviation of the pupil-to-pupil distance of your sample(s) and the standard error of difference between the sample of males and females.
11 Plot a scattergram of pupil-to-pupil distance against age and calculate the degree of correlation.

Questions

a ***Consider the list you have prepared of ways in which humans vary and produce hypotheses to explain why some of the variants you have encountered have survival value, either in modern conditions or in conditions to which the ancestors of your group may have been subjected.***

b ***Was there any doubt about whether any of the characters vary continuously or discontinuously? If so, which characters presented difficulty?***

c ***Do you think it probable that differences in pupil-to-pupil distance are mainly inherent or mainly determined by environment? How could this question be settled?***

INVESTIGATION
17B The influence of environment on development

(*Study guide* 17.2 'The role of inheritance and environment'.)

The character shown by an organism is referred to by geneticists as its phenotype. The genetic constitution which the organism has inherited is its genotype. The two investigations below are intended to show that phenotype results from an interaction of genotype with the environment in which development occurs.

Part A The influence of silver nitrate on fruit flies

Details of the general procedure for keeping and handling the fruit fly *Drosophila melanogaster* are given on pages 38–45 in investigation 17C, 'Patterns of inheritance'. You will use the following true breeding variants:

Name	*Description*
wild type	grey body with darker stripes
yellow body	yellowish–grey body with pale stripes

Procedure

1 Set up cultures of each type of *Drosophila* in specimen tubes on normal fruit fly food medium. Use between 3 and 5 pairs of flies per tube.

2 Set up similar cultures with the same number of pairs of flies but use food medium containing approximately 0.1 % silver nitrate.

3 Put all 4 cultures, carefully labelled, in an incubator at approximately 25 °C.

4 After 8 days remove and dispose of the parent flies. Offspring should emerge about 11 to 14 days after the cultures were started, depending on temperature and other conditions. They are always pale in colour when they first emerge from their pupal case.

5 When the young flies have had enough time to attain their final colour (between 2 and 4 days after emergence) anaesthetize (see pages 41–43) and examine each culture.

6 Prepare a table and record the appearance of the flies of each genotype developed on normal medium and on silver nitrate medium.

Questions

The colour of a fly's body depends on the distribution and quantity of the pigment melanin. Melanin is a product of enzyme-catalysed reactions.

a ***Explain your results in terms of genotype, phenotype, and environment.***

b ***How could both genetic variation and silver nitrate change the concentration of melanin in a fly?***

c ***How could you show that silver nitrate has not changed the genotype of the flies, but merely altered its expression?***

Part B Chlorophyll production in seedlings

You will probably be using a population of tobacco seeds (*Nicotiana* sp.). They were obtained by the self-pollination of a normal green plant from a strain which regularly produces chlorophyll-deficient (albino) seedlings. Albino seedlings cannot photosynthesize and soon die but the condition is inherited and is transmitted by normally green seedlings related to the albino ones.

Procedure

1 Select sufficient dry seeds (they are very small). You will need about 30.

2 Place the dry seeds in a small pouch made from a folded scrap of a closely woven material.

3 Tie the top of the pouch tightly with cotton.

4 Soak the bag of seeds in 1% sodium hypochlorite for two minutes. Make certain the solution has penetrated the bag by squeezing with blunt forceps.

5 Transfer the bag to a large volume of distilled water and agitate thoroughly for three minutes to rinse off the hypochlorite.

6 Open the bag and transfer the seeds with flamed forceps or needle to the agar surface in your dishes. Make sure they are well separated. Do not allow more than 20 seeds per dish or they will be difficult to score.

7 Close the dishes and label them. Leave both dishes at a temperature of 20 °–25 °C, one in the light (but out of direct sunlight) and the other in the dark.

8 After about six days germination should have begun. Inspect each

dish daily (making the inspection brief for the 'dark' dish) and record the colour of the cotyledons of each seed as it becomes apparent. Mark the position of the seedlings with a marker pen on the underside of the dish as they germinate, so that you do not score one twice.

9 When all or most of the seeds have germinated, transfer the 'dark' dish to the same light conditions as the 'light' dish.

10 After 48 hours, record the cotyledon colour once again for the 'dark' dish.

Questions

a ***Explain the presence of colourless cotyledons in both treatments as fully as you can.***

b ***From your observations in this investigation, what general hypothesis can you make regarding the influence of the environment on the phenotypic expression of a genotype?***

INVESTIGATION 17C Patterns of inheritance

(*Study guide* 17.3 'Mendel and his contemporaries', 17.4 'Dihybrid crosses', 17.5 'Autosomal linkage', and 17.6 'A model for the inheritance of continuously varying characters'.)

There are a number of different organisms, both animal and plant, that can be bred easily in the laboratory and can provide the raw material for genetic studies. Ideal organisms will have some or all of the following features:

1 they are simple and cheap to culture
2 pollination can be controlled or sexes can be distinguished before mating can take place
3 they have a short life cycle
4 they take up little space
5 they produce many offspring per mating
6 they show clear, inheritable discontinuous variation
7 pure stocks are readily available.

The number of crosses that you or your teaching group will be able to undertake is likely to be limited by the space and time available. Quickest results are likely to be obtained using fruit flies, *Drosophila melanogaster*, but even with these, an investigation will take four or five weeks to complete. A number of approaches are possible:

1 Your teacher and laboratory technician may set up a cross and allow you to record the results.
2 A small group of students may set up a cross and a whole class may record the results.
3 The class may set up crosses some time before it starts to study inheritance so that results are available to be interpreted when you are ready for them.
4 The class may set up crosses after some understanding of genetics has been gained and the results may be used to extend your understanding.

Different organisms and different types of cross will particularly suit each of these approaches and it is not feasible in the space available to give details of all the numerous possibilities.

A great deal of experimental biology depends upon the maintenance of organisms in good, controlled conditions of culture for long periods of time. Genetic experiments are a good opportunity for you to learn the skills involved in successfully completing a longterm experiment. Whichever organism you work with, you will not succeed without regular and careful attention to the welfare of the animals or plants used, precise labelling of all cultures and containers, and systematic recording of data.

Option A Seeds and seedlings

Flowering plants have rather too long a life cycle to allow a series of crosses to be completed during most school or college courses. However, suppliers have marketed kits containing samples of seed of parental types F_1, F_2, back crosses and test crosses. Some of these have been used in investigations described in this guide but there are many others available. Most kits use characters manifested by seedlings. You will not do any crossing yourself unless you have facilities for growing the seedlings to maturity. Controlled cross-pollination is fairly easy to achieve in tomatoes.

Option B Small mammals

Even if you do not have facilities to keep many animals, a lot of interesting information can be gained by (carefully) recording pedigrees of all small animals such as rats, gerbils, or mice which are bred in the school animal house. It would hardly be worth while to breed many individuals just for genetic study, but they may well be required for studies of growth or behaviour and these purposes can be combined with a longterm genetic study.

If true-breeding stocks of mice can be obtained, and there are facilities for several pairs, you can get a significant number of F_2 and backcross individuals after about two terms, since the life cycle is about 12 weeks. True breeding stocks are not absolutely essential. If a number of coat colour genes are segregating in the existing animal room stock, you can draw up family trees, such as are used by human geneticists (who never have true breeding stock), and can often explain the inheritance of a number of coat colour phenotypes. It may be possible to test hypotheses which explain your mice family trees by buying one or more animals of known genotype and making several test matings. Unless sex linkage is suspected, it is easier to buy in a male for test matings since you can mate it to as many females as you can afford to house and feed. Mammals should always be obtained from an approved source or they may introduce disease to your existing stocks.

Option C Flour beetles (*Tribolium* sp.)

Beetles of two species, *Tribolium castaneum* and *T. confusum*, are often kept in laboratories and have been much used in studies of competition and orientation behaviour. Several true breeding mutant stocks can be obtained. They are extremely easy to culture but suffer from the disadvantage that they are not easy to sex and breed rather slowly (life cycle is six weeks under optimum conditions).

Option D Fungi

A great many fundamental discoveries in genetics have been made using fungi such as *Neurospora crassa*, *Aspergillus* sp., *Sordaria fimicola*, *Saccharomyces cerevisiae*, and *Coprinus lagopus*.

The life cycles of such organisms are rapid and many mutants are known which influence the biochemistry of the cells or mycelium.

Option E Fruit flies (*Drosophila melanogaster*)

The life cycle is illustrated in *figure 13*. The time *Drosophila* takes to complete this depends upon the temperature. Usually we keep experimental cultures at 25 °C, at which temperature the females mate about ten hours after emergence. They will begin laying eggs after a further thirty-six hours and may continue for a period of three to four weeks. Sperm stored since the first insemination may fertilize these eggs; or another insemination may do so. The female can lay over a hundred eggs. A female can only be used once in an experimental cross and she must be unmated (virgin) at the start.

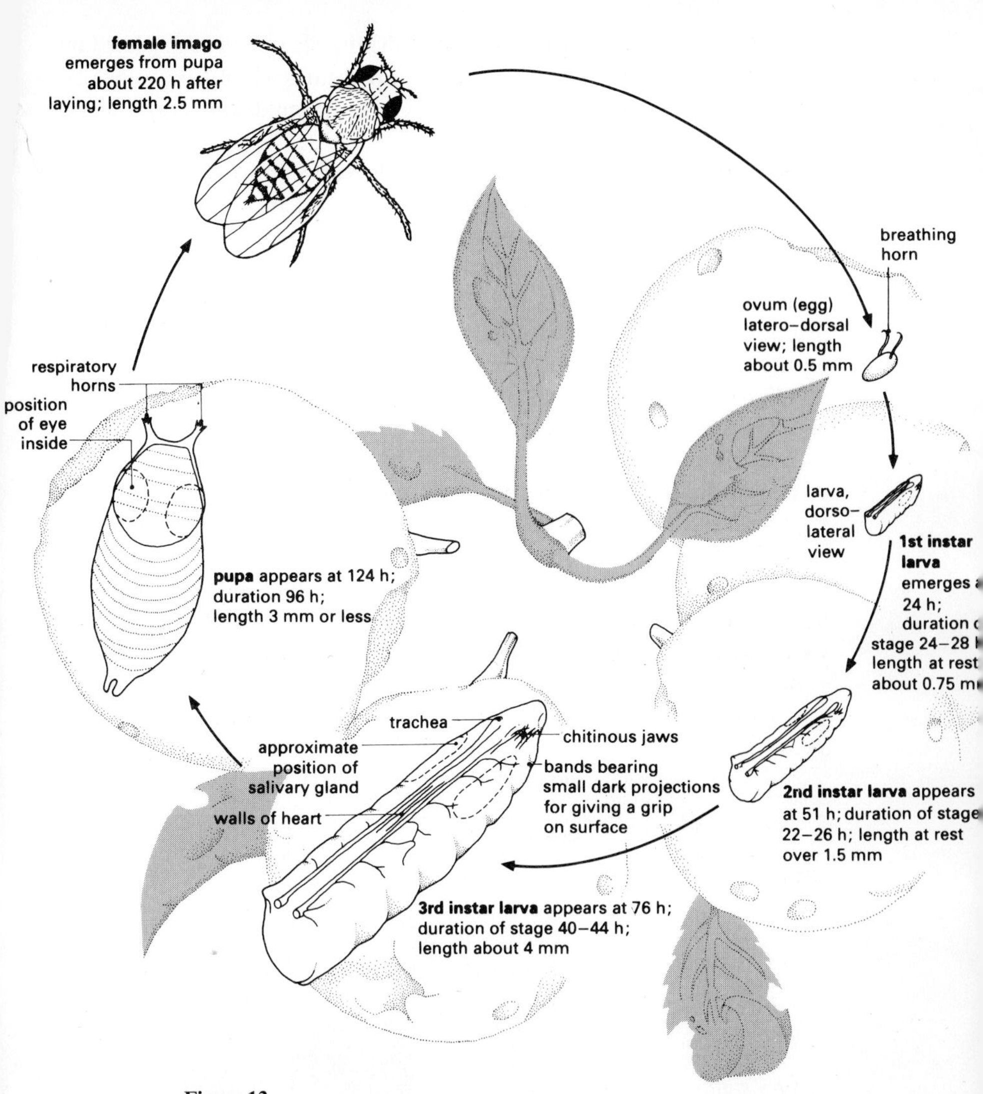

Figure 13
The life cycle of *Drosophila melanogaster*. All sizes and times given are for cultures maintained at 25 °C. Times of appearance are hours after egg-laying (approximate). Until the adult stage this insect lives in and feeds on ripe or rotting fruit covered with yeast cells.

The egg hatches after about twenty-four hours and the larva feeds throughout its five day life, moulting twice. During the third larval stage, the black jaws are particularly obvious and the salivary glands are well developed, with extraordinarily large chromosomes in the cells which are visible only when stained. The larva pupates and an adult emerges from the pupa after a further four days. At 25 °C the life cycle is completed in about eleven days. In normal outdoor temperatures it may take weeks or even months. Temperatures over 30 °C are harmful and may kill or cause the flies to be sterile.

Fruit flies can be cultured on a wide variety of artificial media. Manufacturers market dry preparations which can be poured into a vessel to which a measured amount of water is added, or a porridge-like fresh medium can be made. This will be done for you and will have been poured into culture vessels and left to set. Media contain sugars which promote the growth of yeasts upon which the larvae feed. The adults feed on the sugars. A successful medium must hold sufficient water to allow good growth of yeast for several days without being too wet and sticky. Moulds tend to invade colonies and compete with the yeasts. This is avoided by adding mould inhibitors to the medium and sprinkling dried yeast powder on the surface just before introducing the flies.

Large cultures are set up in milk bottles. These are better than flasks as they can be banged down hard on a soft surface such as a book, to dislodge flies from the bung, without breakage. Small cultures are set up in specimen tubes.

A culture tube ready for flies to be introduced is shown in *figure 14*.

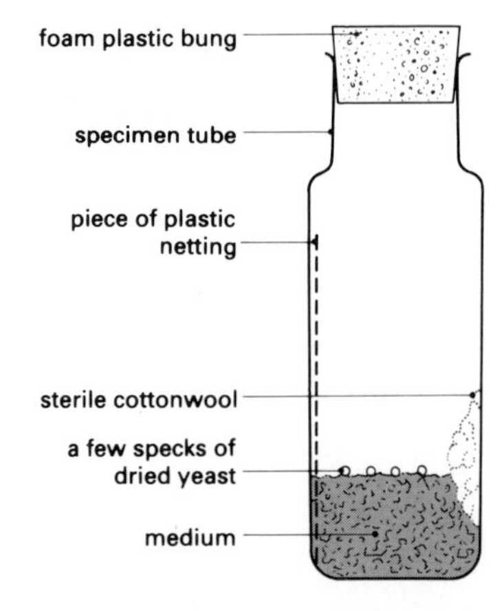

Figure 14
The specimen culture tube for *Drosophila melanogaster*, prepared.

A specimen tube culture should be set up with between two and four females and a similar number of males. A milk bottle culture should have between ten and twenty females. If the colony is too small, the growth of moulds and the breeding of mites are encouraged.

In order to handle fruit flies, you must anaesthetize them.

Procedure

1 Obtain an etherizer (see *figure 15*) and make sure it is clean, dry, and free of flies.

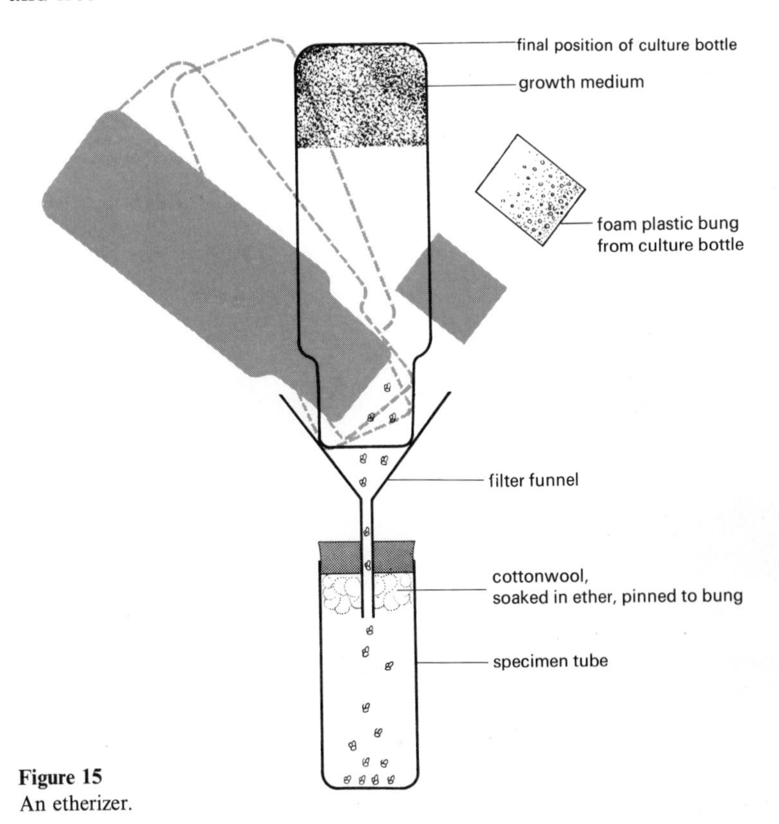

Figure 15
An etherizer.

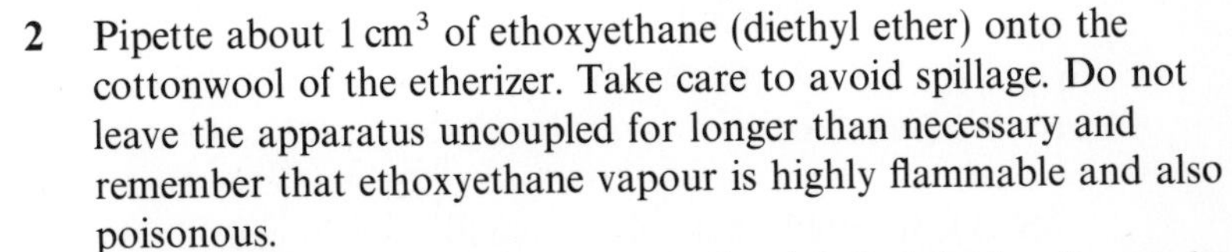

2 Pipette about 1 cm^3 of ethoxyethane (diethyl ether) onto the cottonwool of the etherizer. Take care to avoid spillage. Do not leave the apparatus uncoupled for longer than necessary and remember that ethoxyethane vapour is highly flammable and also poisonous.

3 Have ready all the apparatus, including labelled fresh culture tubes, that you will need in order to examine and handle the flies. You will not have time to look for things once the flies are anaesthetised.

4 *Either*
Bang the upright culture tube or bottle sharply down to shake the flies off the bung and to the bottom of the vessel. Quickly remove the bung and invert the culture over the etherizer as in *figure 15*.
Or
If the medium in the culture is very wet and likely to flow if the

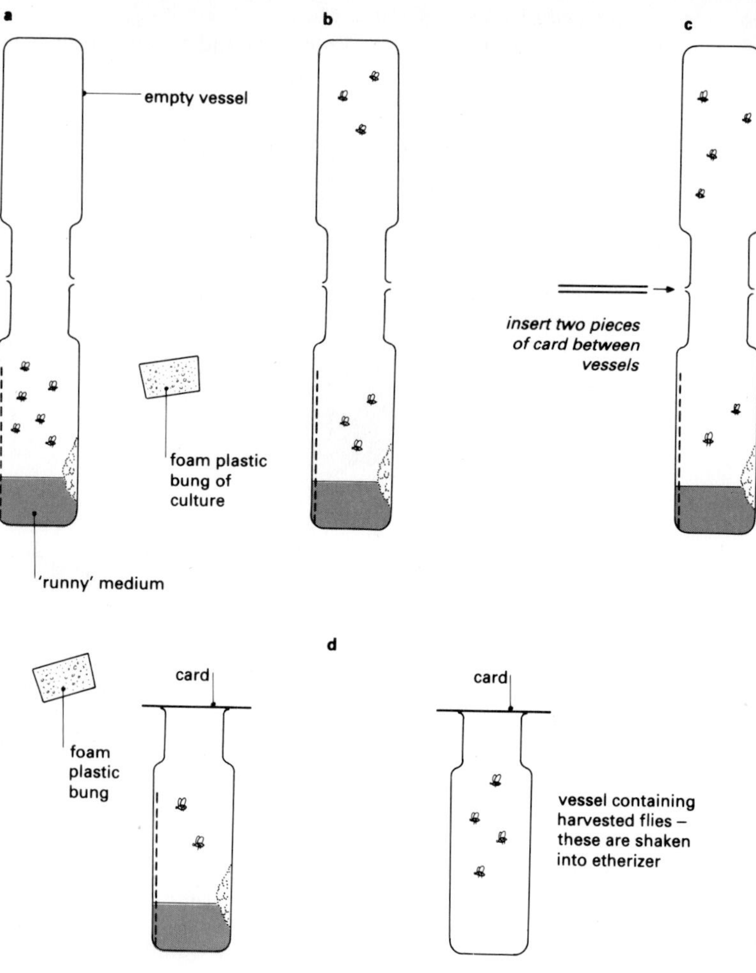

Figure 16
Harvesting the flies.

vessel is inverted, shake down the flies as above and remove the bung. Invert a clean, dry empty bottle or tube over the culture as in *figure 16* and 'harvest' the flies in this.

5 Tap and shake the flies down into the tube of the etherizer.
6 Remove the culture or harvest vessel and replace its bung. Watch the flies carefully and as soon as the last one stops moving or clinging to the surface of the tube, open the etherizer and empty the flies out onto a white tile. Avoid doing this in a draught.
7 Count, sex, and score the phenotype of the flies on the white tile, moving them gently into piles with a small, soft paintbrush.

8 Set up fresh cultures or crosses by sweeping the required flies onto the inner side of the new culture vessel, pushing in the bung, and leaving it on its side until they recover and become fully active. If an inactive fly touches the medium at the bottom of a tube (or glassware wet with condensation), it will stick and die.

9 Dispose of unwanted flies in a jar of 70 % ethanol.

Further notes on etherization

1 If liquid ether touches a fly, it will die.

2 Prolonged exposure to ether vapour sterilizes flies. Make sure that fresh tubes are free of it.

3 Flies killed by ether have their wings erect over their backs as if in flight. Live flies fold the wings horizontally on their backs.

4 If a fly recovers from the ether (ignore slight twitches) while you are scoring, sweep it back into your etherizer through the funnel and wait until it is still again.

Distinguishing sexes

The best method of distinguishing the sexes is to examine the external genitalia on the ventral surface of the abdomen. Females have pointed abdomens and the genital papilla is visible from above. Males have a dark brown genital area on the ventral surface of the abdomen which is a characteristic 'square' shape and is darkly pigmented even in newly hatched flies. It is the most obvious feature of the males and is clearly visible to the naked eye when the fly is lying on its back (*figure 17*).

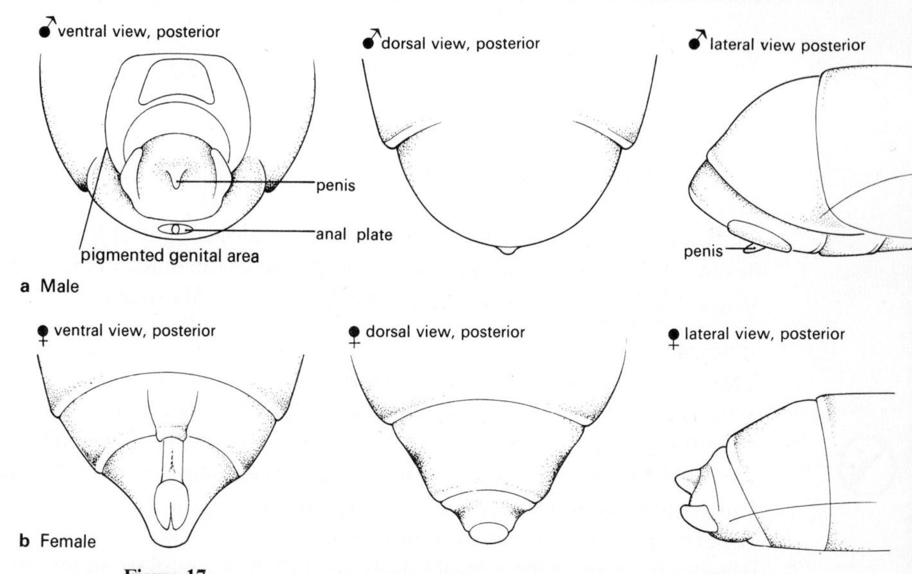

Figure 17
Male and female *Drosophila*: ventral, dorsal, and lateral views of the genitalia.

Take care when you are examining the abdomens of dead flies and those which have been crushed or starved because they tend to become shrunken and females then look like males.

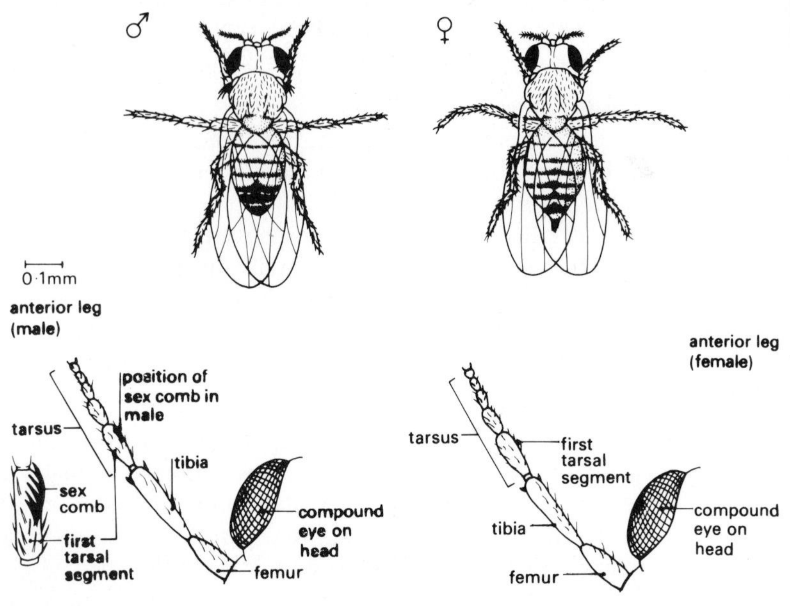

Figure 18
Other distinguishing characters of male and female *Drosophila*.

You can also use the following characters of males to confirm your findings.

They have sex combs on the anterior pair of legs and broader, blacker stripes near the tip of the abdomen, a rounded abdomen, and a smaller overall size than the females (*figure 18*).

In the females the abdomen is clearly striped throughout the length of the body when viewed from above whereas, in the male, the last few stripes tend to merge to give a jet black tip to the abdomen.

Obtaining virgin females

1 Leave the stock culture tube in a cool place (for example, the windowsill of the laboratory) overnight. 5–10 °C is suitable.

2 First thing in the morning, etherize all the adults in the stock tube. This is the only occasion when it is necessary to remove every fly. (If you require male flies, these may be isolated in a fresh tube of medium at once, since male virgins are not essential.)

3 Leave the culture tube free of adults with its bung in place in an incubator at around 25 °C.

CHAPTER 18 # THE NATURE OF GENETIC MATERIAL

INVESTIGATION
18A Testing for DNA using the Feulgen technique

(*Study guide* 18.1 'The search for genetic material'.)

Under conditions of mild acid hydrolysis, DNA loses some of its purine bases from their attachment to the first carbon atoms of the deoxyribose sugar molecules. This allows the DNA to react with the Feulgen stain and produce a purple colour. These reactions do not happen with RNA because ribose sugar molecules have a hydroxyl group on their second carbon atom. (See *Study guide II*, figure 74.)

Deoxyribose is a reducing sugar and the colour developed during staining depends on the reduction of a dye which is colourless in its bleached (oxidized) state.

The Feulgen technique gives a useful diagnostic test for the presence of DNA in a cell. DNA and only DNA stains purple while the remainder of the cell is unstained.

Procedure

1 Any fresh tissue likely to contain cells undergoing mitosis or meiosis is usable. It must be fixed in freshly made ethanoic alcohol for a minimum period (see page 6).

2 Warm 5 cm^3 of molar hydrochloric acid to 60 °C in a water bath. Rinse the material in distilled water in a watch-glass and then place it in the preheated acid at 60 °C for 7 minutes.

4 Rinse the material in water once again and then immerse it in Schiff's reagent for 30 minutes or longer if possible, in the dark. When the material is stained a dense purple it is ready to be taken on to the next step.

5 Squash the preparation using one of the methods on page 7.

6 Examine the preparation under L.P. and H.P. and identify as accurately as possible any parts of the material that are stained.

Question

a ***From the evidence of your preparation, what conclusions do you come to about the DNA content and distribution in the structures of 1 the nucleus and 2 the cytoplasm?***

4 Within the next 10 hours, anaesthetize all the adults that have emerged. Any females found will be virgins.
5 Put the virgin females into a fresh tube of culture medium on their own, as you did for the males, until they are needed for setting up a cross; alternatively, set up the crosses immediately.
6 If F_1 flies are to be crossed *inter se* to get an F_2 generation, virgin females are unnecessary.
7 Virgin flies are often very pale and difficult to sex by looking with the naked eye at their abdominal stripes and shape. Use a dissecting microscope and look for the sex comb and for the difference between male and female genital apparatus (see *figure 17*). Correct sexing at this stage is absolutely vital.

Scoring results

1 Remove and dispose of the parental flies 8 days after setting up any cross. They must not be confused with their offspring.
2 At 25 °C, the offspring should emerge as adults 11–14 days after the culture was set up. Remove, score, and dispose of the offspring every 2 days. If they become crowded some will die and this will not be a random sample. The ratio of phenotypes in the offspring will be distorted by selection.
3 It is important to carry on harvesting and scoring offspring until no more emerge as some phenotypes (for example, vestigial wing) are slower to develop than others.
4 Do not score flies emerging more than 19 days after setting up the culture as these may be of the next generation, that is, the offspring of those being scored.
5 Record your results in a table such as table 1. It should contain columns for each phenotype character in each sex and sufficient rows for one row to be used on each occasion that the flies are scored.

	Wild type wings		*Vestigial wings*	
DATE	Female	Male	Female	Male
12th June	||	|||	|	|
13th June	~~||||~~ |	||	|||	|
Totals				

Table 1

Handling the data

1. When all crosses are completed, tabulate all the totals for each phenotype of offspring for each type of cross.
2. Examine your table and formulate a hypothesis to explain the data, for example, that a dihybrid segregation is involved and that there may be autosomal linkage. State the symbols you are using for genes and their alleles.
3. Predict the expected ratios of offspring in F_2 and backcross/test cross generations on the basis of your hypothesis.
4. Work out expected numbers of offspring, given the total number actually scored and your hypothetical ratios.
5. Carry out a χ^2 test to determine whether the actual numbers of various phenotypes observed conform to the expected numbers on the basis of your hypothesis. Modify your hypothesis if necessary.
6. Suggest further crosses which might test your hypothesis.

INVESTIGATION
17D Linkage and linkage mapping in tomato

(*Study guide* 17.5 'Autosomal linkage'.)

In this investigation the inheritance of three genes is considered in a trihybrid cross. The loci involved are on the same chromosome and therefore exhibit linkage and you will be able to calculate cross-over values and work out the relative positions of the three loci on the chromosome.

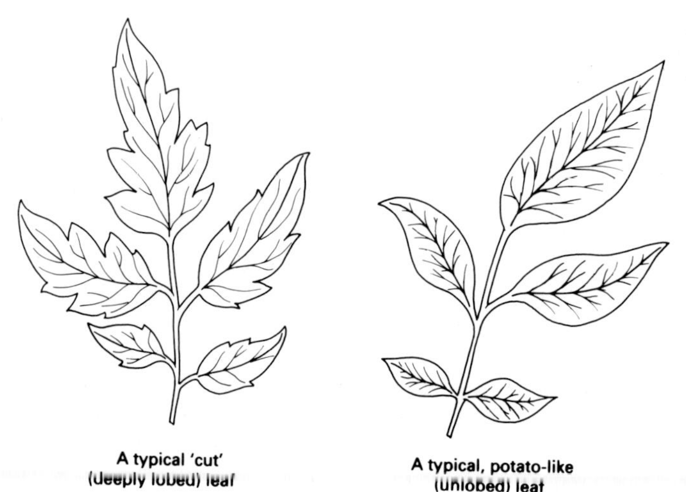

Figure 19
Lobed and unlobed leaves.

The organism concerned is tomato, *Lycopersicon esculentum*. The necessary crosses have been made for you (by removing stamens from an immature flower in this self-pollinating species and then dusting the stigma with appropriate pollen). You will be provided with families of seeds and it is important to sow them with great care since a high germination rate is important if meaningful results are to be obtained.

The parent P_1 had purple stems, normal green foliage, and deeply lobed (tomato-like) leaflets (genotype BB SS FF). The parent P_2 had green stems, yellowing leaves, and non-lobed (potato-like) leaflets (genotype bb ss ff).

You should sow the seed of both the F_2 and the test cross (which is a backcross to P_2) and also a few seeds of the two parental types so that you will be able to check that you have properly observed the contrasting characters to be studied.

The three genes are listed below, together with a brief description of the phenotypes produced by their two alleles.

Gene controlling:	*Allele symbol and phenotype produced by*	
	the dominant allele	*the recessive allele*
stem pigmentation	B (purple)	b (green)
leaf colour	S (green)	s (yellowing)
leaf shape (see *figure 19*)	F (deeply lobed margin)	f (non-lobed margin)

Procedure

1. Fill seed trays approximately three-quarters full with potting compost which has been slightly dampened, and pack the compost down.
2. Plant the seeds 2 cm apart in rows that are spaced so that there is approximately 2 cm between the rows.
3. Cover the seeds with 5 mm of compost and press gently down.
4. The seed trays should then be covered with a dark cover (such as plastic or glass lid covered with newspaper) and left at 20 °C.
5. When the first signs of germination are noticed, the dark cover must be removed. The seedlings must now be in a brightly illuminated position. If a transparent cover is left in place great care must be taken that seedlings are not overwatered or left in hot sunshine or they will suffer fungal infection. If no transparent cover is used equal care is needed to see that the compost does not dry out.
6. Leave the seedlings to grow until approximately 14 days after planting. Once the first true leaves have appeared and have expanded a little, the seedlings may be 'scored' for all three

phenotypic characters. In order that the stem pigmentation should be as clear as possible, the seedlings need to be placed in cool, brightly lit, fairly dry conditions for 36–48 hours before scoring is done.

7 Sorting is best done by harvesting the plants and dividing them successively into groups according to each characteristic. You will end up with eight groups of plants.

8 Record the frequency of each of the eight phenotypes of plant you now have.

9 From your knowledge of genetics, predict the outcome of the cross which produced the seeds that you planted, assuming that the three genes are unlinked. This will give you the 'expected' results for use in the chi-squared test.

10 You are now in a position to form a null hypothesis and test it using the chi-squared test.

11 Interpret the results of the chi-squared test, and use the ideas of linkage to explain any departure from the expected result that was revealed by the test.

12 Use the results of the cross to determine the cross-over value (COV) for each gene pair in this cross, where COV is given by

$$\text{COV} = \frac{\text{number of appropriate recombinant types}}{\text{total number of offspring}}$$

13 Express each COV as a percentage.

Questions

a ***Put into words the null hypothesis that your chi-squared test was evaluating.***

b ***State 1 the value of chi-squared given by your results, and 2 the probability of obtaining that chi-squared value. Put into words the outcome of your chi-squared test.***

c ***Why was it not necessary to test the results of the cross against 'expected' results, which assumed that the genes were completely linked with no crossing over?***

d ***Use the results of your cross to draw a linkage map of these three genes, which are all carried on chromosome III; assume that the locus of the gene for stem colour is 74 map units from the end of the chromosome. Explain how you arrived at the map that you have drawn.***

e *Suppose the investigation were to be repeated starting with homozygous parents:*
P_1 green stem, yellowing leaves, deeply lobed leaflets (bb ss FF)
P_2 purple stem, green leaves, non-lobed leaflets (BB SS ff).
1 Would the linkage between the genes be the same?
2 What cross would be needed to work out the cross-over values required to map the loci?

f *Briefly outline the advantages and disadvantages for a species in possessing genes which are 1 unlinked, 2 loosely linked, and 3 tightly linked.*

INVESTIGATION
17E Mutation in yeast

(*Study guide* 17.7 'Mutation'.)

Ordinary bakers' yeast (*Saccharomyces cerevisiae*) is a facultative anaerobe. It usually respires aerobically but if oxygen is unavailable it can respire anaerobically. Occasionally respiratory-deficient (RD) forms arise by mutation which lack some of the enzymes necessary to enable them to respire aerobically; they are obligate anaerobes, even in the presence of oxygen. These RD forms can be identified if grown in the presence of triphenyl tetrazolium chloride (TTC) which is taken up only by aerobically respiring cells and reduced to a deep red colour. The cells which are respiring anaerobically do not reduce the TTC and remain white.

If a suitably dilute suspension of yeast cells is spread on the surface of an agar-containing food medium they will grow as discrete colonies, each colony being the result of asexual reproduction (by budding) of one initial yeast cell. Any visible differences between colonies grown under identical environmental conditions most probably represents a genetic difference between the single cells that initiated those colonies. In this investigation you are going to identify the RD colonies as evidence of spontaneous mutation in a population of yeast.

Procedure

1 Using flamed forceps, and taking any necessary sterile precautions, select a granule of dried yeast that is about 1 mm in diameter, and put it into a stoppered bottle containing 10 cm^3 sterile distilled water.
2 Shake the bottle until the yeast is completely dispersed in the water.
3 Draw up the suspension into a sterile syringe. Draw a little air in and shake the syringe to mix the suspension further.

4 Expel all but 1 cm^3 of the suspension and draw up 9 cm^3 of sterile water once again and mix.
5 Repeat step **4** once more.
6 Place one drop of this diluted suspension onto the surface of each of the plates of agar medium and quickly spread the drop over the entire surface of the agar with a sterile glass spreader.
7 Close the Petri dish, seal in two places with strips of sticky tape, and label each dish with your name and the date. Invert the dishes and incubate at 30 °C for 48 hours.
8 Melt a tube of agar containing glucose and TTC in boiling water and transfer it to a water bath at 40 °C to cool for 10 minutes.
9 Pour a thin layer of the TTC-agar over the surface of your plates which now have yeast colonies on them and leave the new layer to set.
10 Incubate the plates at 30 °C for between 2 and 48 hours more.
11 Examine the plates and the colonies carefully. It may be helpful to use a stereomicroscope or hand lens. Note the colour, size, and shape differences between colonies on any plate, and record your observations.
12 Count and record the number of entirely white colonies and the number of non-white or mixed colonies in each plate.

Questions

a ***Explain why the occurrence of white colonies can be taken to be the result of spontaneous mutation and not, for example, the influence of some environmental factor on existing mutant cells, or of contamination.***

b ***Explain any observations that you made in step 12 above.***

c ***Why does the technique employed here only reveal the respiratory-deficient mutants? Are there any other mutant yeast colonies present on your plates?***

INVESTIGATION
17F The effects of irradiation in plant seeds

(*Study guide* 17.7 'Mutation'.)

You will be supplied with tomato seeds produced as a result of self-fertilization by plants heterozygous for the gene Xa–2 (xanthophyllic–2) and its wild type allele $Xa–2^{+}$. Xa–2 causes chlorosis (a reduction in the amount of chlorophyll produced by the plant). There is no dominance, with the result that selfing a heterozygote produces a

1:2:1 ratio of phenotypes. Homozygous xanthophyllic individuals are yellowish white (highly chlorosed) and die soon after germination. Homozygous wild type are of course a normal green colour. The heterozygotes appear distinctly yellow and are slower growing than their normal counterparts.

Procedure

1. Sow each packet of seed in a separate seed tray following the method outlined in investigation 17D, 'Linkage and linkage mapping in tomato', steps **1**–**5**. Count and record the number of seeds sown in each tray.
2. Label each tray with the radiation dose received.
3. As soon as germination is complete (some seeds may not germinate), record the numbers of green, yellowish–green, and yellowish–white seedlings in each tray.
4. While scoring the seedlings look out for obvious departures from normal phenotype and take note of these. Patches of green or of white on the yellowish–green (heterozygous) seedlings are the most likely changes. Record the number of such patches in each tray.
5. Remove the green and the white seedlings (the homozygotes) by cutting them off at soil surface with a sharp pair of scissors and picking them out with forceps. Examine each carefully for abnormalities and discard it.
6. Leave the heterozygotes to grow until the second true leaf pair has developed noticeably. Development will possibly reveal more patches of green or of white tissue. Add these to your table of data but be careful not to score any patch more than once.
7. After the second true leaf has developed in the control tray (zero radiation dose), harvest the plants. Give the entire pot or tray of seedlings a few days for the compost to become very dry (but not so dry that the seedlings wilt). Then very carefully empty the tray onto newspaper and shake the seedlings, supporting their root network, until all individuals are separated from one another.
8. Hold each seedling by its shoot and its roots and gently shake or brush with a paint brush until all particles of compost that are clinging to the roots are removed.
9. Measure the full length of the entire seedling, from the tip of the longest leaf to the tip of the longest root, by placing the seedling on to millimetre graph paper, or alongside a ruler. At the same time, record separately the length of the root and shoot portions of the seedling.
10. Find the mass of the seedlings to the nearest 10 milligrams.
11. Calculate the mean values of the lengths and masses of your

seedlings and, if you have sufficient replicates, determine the standard deviation of each variable in each treatment.

12 Plot the means and standard deviation of each variable that you record against radiation dosage to show the overall trend in the response to irradiation by these seeds.

13 Statistical analysis may be undertaken as well as the simpler graphical analysis:
a the effects of two different doses of radiation may be compared, using the standard error of difference test (provided there are enough replicates), or **b** the correlation between radiation dose and growth may be assessed using Spearmann's rank correlation test.

Questions

a ***Put forward and explain a simple hypothesis to account for the occurrence of chlorosed or green patches of tissue in the leaves of a heterozygous plant.***

b ***An organism which possesses cells or tissues whose genotypes differ one from another is called a mosaic. What evidence do you have that enables you to argue that some of the irradiated seedlings are mosaics? What additional evidence would give added firm support to your argument?***

c ***Attempt to explain the effect of radiation dose on the germination, survival, and growth of the seeds and seedlings.***

INVESTIGATION
18B Testing for DNA and RNA using methyl green pyronin

(*Study guide* 18.6 'The breaking of the genetic code'; Study item 18.63 'The message linking nucleus and cytoplasm'.)

The use of Feulgen stain after the hydrolysis of a tissue (investigation 18A) makes it possible to locate DNA within the cells but the reagent does not detect RNA so another method is needed to show its whereabouts. The use of the mixed stain, methyl green pyronin, allows the location of RNA as the stain colours both types of nucleic acid.

Procedure

1 Take an onion or any similar bulb and cut a small square (sides of about 5 mm in length) in the organ to a depth of about 4 mm. Using a mounted needle, remove the block of tissue.

2 Separate the pieces of fleshy leaf base from each other and, with a mounted needle, carefully scrape away at the corner of the concave surface of one of the leaf bases. A very thin, transparent layer of tissue will be apparent on this inner surface. Remove this layer, which is only one cell thick, with a pair of forceps.

3 Immediately fix this piece of epidermis for at least 5 minutes in absolute ethanol.

4 Transfer the piece of tissue to a small quantity of the stain in a watch-glass or small beaker for 10 to 12 minutes.

5 Pick the tissue out of the stain and wash it thoroughly in a beaker of tap water. It may be necessary to wash it in two changes of water if a lot of stain is transferred with the tissue.

6 Mount the tissue in a drop of water on a slide and carefully flatten out any folds in it with a pair of mounted needles. Examine the tissue carefully under L.P. and H.P.

7 Draw a representative cell under H.P., paying special attention to the location of *any* coloured material. The stain will colour DNA a blue–green colour and RNA pink. Label your drawing fully.

Question

a ***Using data from this and from the previous investigation, make a careful statement about the distribution of the two nucleic acids in the cell.***

INVESTIGATION

18C Making a model to illustrate the chemical nature of a gene

(*Study guide* 18.4 'The chemical structure of DNA'; 18.5 'A model of the DNA molecule'; 18.6 'The breaking of the genetic code'; 18.7 'The synthesis of RNA'.)

The nature of the nucleic acids DNA and RNA, and their relationship to enzymes and other proteins, have formed one of the most important areas of scientific study since the 1940s. Currently accepted hypotheses about the functions of the nucleic acids have been outlined in *Study guide II*, pages 115–21, and can be found in a great many biological and biochemical texts. It is important that you are thoroughly familiar with these hypotheses, so that you can appreciate the many implications that they have for almost all branches of biology. It may help your understanding if you can construct or investigate models of these molecules and their operation. You should be careful to realize that such models are limited in their application and do not necessarily always give a completely helpful representation. *Figure 20* shows a model built to accurate dimensions by the discoverers of the structure of DNA.

Procedure

1. Find a diagram which shows the structure of the four deoxyribonucleotide phosphate sub-units which make up a DNA molecule (as in figure 68 of *Study guide II*.)
2. Draw simplified representations of each of these molecules on stiff card. The drawings must be to the same scale.
3. Cut out the four diagrams with a knife or scissors.
4. Place each model in turn on paper and draw around it to reproduce its outline. Repeat this about twenty times and then cut out the outlines so that you have a good supply of the four nucleotide phosphate models.
5. Label the bases on the models A, T, G, and C as appropriate.
6. Join a row of the models together with sticky tape to represent a single strand of DNA. The models should be joined so that they form a vertical series, with the phosphate of one model joined to the deoxyribose portion of the model above it. All the bases should now be aligned on one side of the DNA strand.
7. Using the principles of base pairing, produce a complementary strand of DNA which will match your first model strand, and make a section of double-stranded DNA.
8. Make your model illustrate replication, by separating the two sides of the double DNA strand and pairing the bases on each side with those of additional deoxyribonucleotides.

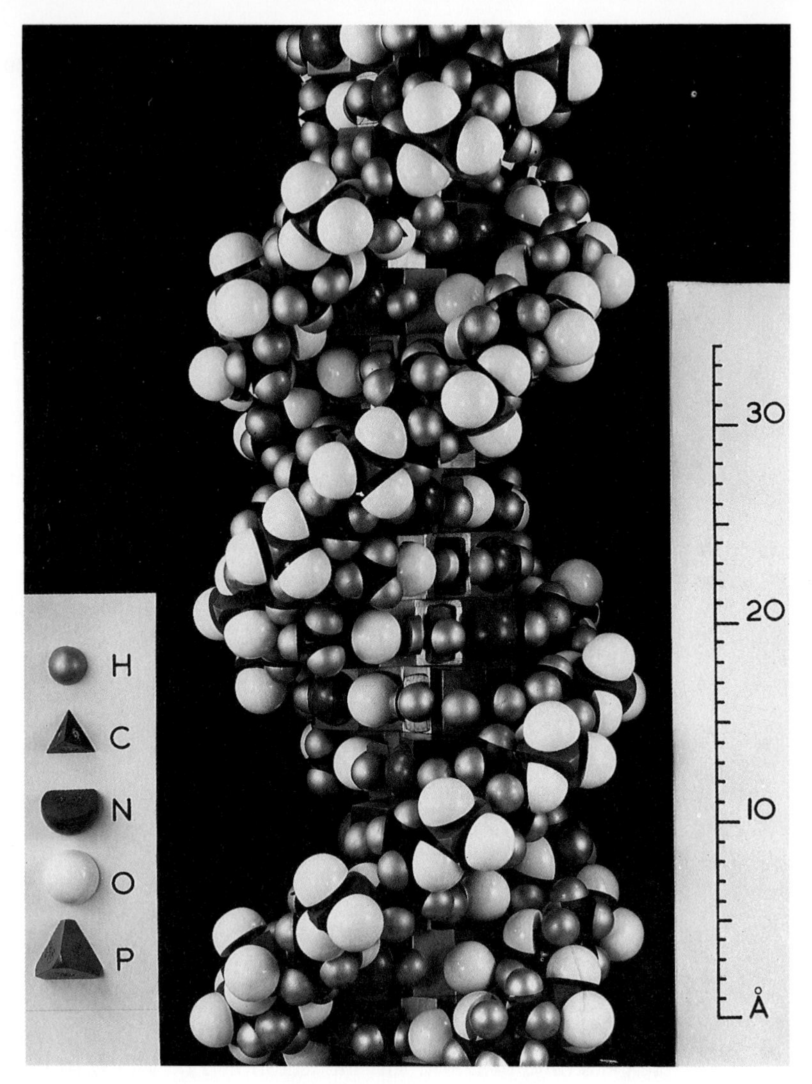

Figure 20
A model of DNA.
Photograph, Department of Biophysics, King's College London (KQC).

9 Prepare a supply of ribonucleotide models, using your original card outlines but differently coloured paper from that used for the DNA sub-units.

10 Use these ribonucleotide models and a single strand of the DNA molecule to construct a matching messenger-RNA model for your DNA.

11 Devise for yourself paper models of ribosomes, transfer RNA molecules, and amino acid molecules and use them to work through the mechanism of DNA transcription and translation with fellow students; discuss the limitations of the paper model.

12 Find someone who is unfamiliar with current ideas about the mechanism of protein synthesis and use your model to explain current ideas to him or her.

CHAPTER 19 GENE ACTION

INVESTIGATION
19A Dwarfism in peas

(*Study guide* 19.3 'Control of transcription in higher organisms'–see 'Gene activity in polytene chromosomes'; and 19.5 'Genes that influence metabolic reactions in Humans'–see Study item 19.51 'The influence of thiamine on mutant tomatoes; an analysis of an experiment'.)

Dwarfism is an inherited condition in peas. Mendel demonstrated that it is produced by a recessive allele (t) which causes the homozygotes (tt) to be about 50 cm high. The heterozygotes (Tt) and the homozygous dominants (TT) are much taller, over 150 cm tall. The modal height of each true breeding variety of pea is different because several genes besides T/t influence this character. Most modern commercial varieties are dwarf, and really tall varieties over 200 cm tall are difficult to get because gardeners prefer dwarfs. (See *figure 21*.)

Growth in peas involves elongation of the stem and this is influenced by a number of plant hormones including IAA and gibberellic acid. It is a reasonable hypothesis that the T/t locus in some way influences the amount of a hormone which a plant synthesizes and so affects stem elongation.

Many factors will influence growth, including light intensity, competition, the size of the container used, and the amount of water available. Peas very quickly die of a fungal infection if overwatered while germinating. After reading the instructions that follow, plan within your group how these factors are to be kept as constant as possible and how the plants are to be looked after. Some of the ways in which hormones influence plant growth are discussed in *Study guide II*, Chapter 24, and *Practical guide 6* investigates a possible mechanism for the action of gibberellic acid in seed germination.

Procedure

1 Sterilize the surfaces of several seeds of a tall variety of pea with 1% sodium hypochlorite solution. Soak them for 24 hours in a dish of clean water.
2 Label several containers of standard size, *e.g.* plastic drinking cups (with drainage holes), and fill each with a similar quantity of potting compost. Leave about 3 cm clear at the top of each pot to allow watering without spillage.
3 Push sticks into the compost of each pot to support the seedlings as they grow.

Figure 21
Tall and dwarf peas. These plants are genetically different. Hormone treatment allows the dwarf to resemble the tall in phenotype.
Photograph, Duncan Fraser.

4 Plant an appropriate number of soaked peas in each pot. Two or three would be suitable for a plastic drinking cup.

5 Soak the compost with water, allow to drain well, and set aside to germinate.

6 Repeat steps **1–5** with a dwarf variety of pea.

7 As soon as green shoots have appeared, sort out the pots, rejecting those where germination has failed. You should aim to have the same number of healthy seedlings in each pot and may have to transplant some seedlings to fill in gaps.

8 Sort each variety of plants into three equal groups in which the size distribution is similar. You should have three groups of tall plants and three groups of dwarfs.

9 For each variety label one group of pots 'IAA', another group 'Gibberellic acid' and a third group 'Control'.

10 Treat each plant of each group with the appropriate solution. Apply one drop to the apical bud. You are provided with solutions of 30 p.p.m. gibberellic acid and 30 p.p.m. IAA.
You may prefer to spray the plants lightly with the solution, in which case use a 10 p.p.m. solution of each substance and make sure that each group is sprayed quite separately. (An aerosol of the spray could drift several metres in a draught).

11 Repeat the treatments at least once per week for three or four weeks, or until at least one of the treatments has had a visible influence on growth.

12 Assess the growth of each plant as follows:
1 Count the number of expanded leaves on each plant.
2 Count the number of visible internodes (do not forget those between scale leaves near the soil).
3 Measure the plants from soil surface to the apical bud.
(This will be below the first leaf and protected by a pair of bracts at the base of the first leaf.)

13 Tabulate all the data.

14 Calculate the mean height, the mean leaf number, and the mean number of nodes for each variety and each treatment.

Questions

a ***In what way precisely do the untreated (control) plants of each variety differ in phenotype?***

b ***What effect or effects on growth have the two plant hormones used had on each variety?***

c ***Can we assume that the allele t blocks the synthesis of one or both of the hormones?***

d ***Suggest other investigations which might be performed to investigate further the situation revealed by your results.***

e ***It might be supposed that farmers would lose yield by growing dwarf varieties of plants such as peas or wheat. In fact the yield of seeds from a field of dwarf and of tall plants is similar or the dwarf crops yield may be greater. Do your observations suggest a reason for this?***

INVESTIGATION
19B The biochemistry of cyanogenesis in *Trifolium repens* (white clover)

(*Study guide* 19.1 'Mutant complementation in *Neurospora*' – see Study item 19.11 'Mutations and metabolic pathways'; and 19.6 'Gene expression in heterozygotes'.)

Plants of *Trifolium repens* (white clover) are often able to release hydrogen cyanide from their leaves. This only occurs when the leaf tissue is damaged in some way. The cyanide is produced by the breakdown of a substrate (a cyanide-containing glucoside) to give glucose, a ketone, and hydrogen cyanide. This happens very slowly *in vitro* at normal temperatures, but an enzyme (a glucosidase) can speed it up greatly. The general reaction for one particular glucoside is shown in *figure 22*.

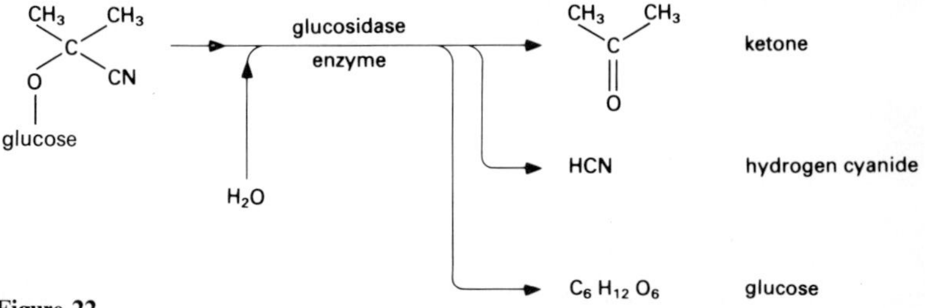

Figure 22
The conversion of a glucoside to hydrogen cyanide in *Trifolium repens.*

It seems that in a normal leaf cell the enzyme and the substrate are kept apart, though we do not know exactly where each is located. In order for cyanide to be released in significant quantities:

1 both enzyme and substrate need to be present
2 the enzyme and substrate need to be brought together.

In the absence of the substrate clearly no cyanide could be liberated; in the absence of the enzyme, a plant which contains the substrate could demonstrate slow cyanide liberation by the spontaneous breakdown of the glucoside.

As hydrogen cyanide is highly poisonous, two questions arise and we will take these up in investigation 20B, 'Cyanogenesis and selection in *Trifolium repens*'.

1 Does a leaf which produces HCN inhibit its own metabolism or growth?
2 Does the ability to produce HCN deter predators which might otherwise eat the leaves of the plant?

The plants grow and reproduce by stolons which creep over the surface of the soil or among the stems of grasses and root at the nodes. Each leaf is divided into three leaflets. In the instructions that follow, it is assumed that leaves are of average size. If a particular plant has very large or very small leaves, then vary the number tested accordingly. Only young, healthy non-wilted leaves are worth testing. You may be provided with a collection of plants growing in pots or trays or asked to collect some from a neighbouring field. If the latter is the case, make arrangements to mark the clumps from which you collected the material so that they may be visited later.

Procedure A: identification of cyanogenic and acyanogenic plants from a mixed population

1 Pick two young leaves from near the tip of a stolon of the plant to be tested. Place them immediately in a small stoppered tube.
2 Push them to the bottom of the tube with a clean glass rod.
3 Taking care not to wet the sides of the tube, add 1 drop of distilled water and 1 drop of methylbenzene (*CAUTION:* do not inhale this substance).
4 Bruise and crush the leaves with the glass rod, taking care not to contaminate the sides of the tube near the top and not to break the tube.

5 Hold a piece of sodium picrate paper with the clean dry forceps inside the mouth of the tube and trap it with the stopper so that it hangs in the atmosphere of the tube but does not touch the wet areas of the tube or the crushed leaves.
N.B. Sodium picrate is poisonous. Do not touch the papers; wash your hands after using them.
6 Label each tube with your name and the source of clover used.
7 Set up a control tube with your experimental tubes.
8 Place all tubes in an incubator at between 25 °C and 35 °C. It is a good idea to put an elastic band round several tubes or put them in a box or beaker so that they cannot fall over.

9 After about two hours and again after about twenty-four hours, examine the tubes and record the colour of the papers in each. The results for procedure A may be classified by reference to table 2.

Result of the picrate test	*Cyanogenic glucoside*	*Enzyme breaking down glucoside*	*Description of plant*
Brown after 2–3 hours at 30 °C	present	present	strongly cyanogenic
Orange or very pale brown after 24 hours	present	absent	weakly cyanogenic
Yellow after 24 hours	absent	unknown	acyanogenic

Table 2
Classification of cyanogenesis in *Trifolium repens.*

Procedure B: distinguishing between two types of acyanogenic plant

1 Heat some strongly cyanogenic leaves in a glass tube in an autoclave for at least 30 minutes. Remove the tube and empty out the leaves.

2 Take three small, stoppered tubes. Into one of them put two autoclaved leaves; into another put one acyanogenic leaf for testing and one of the autoclaved leaves; into the third tube put two acyanogenic leaves. All three acyanogenic leaves must be from the same plant.

3 Carry out steps **2–9** of procedure A on each of the three tubes. Crush and mix the two leaves together thoroughly in step **4**. Do not forget to include a control.

4 Gather all the results of procedures A and B for your class and record them in a table.

Questions

a ***Describe carefully the function of each of the tubes you set up in B, procedure 2.***

Assume that the manufacture of glucoside by the clover depends on the presence of an allele (G) with a recessive, non-productive allele (g) and that the enzyme synthesis is similarly controlled but by a different locus, with a dominant allele (E) producing the enzyme and a recessive allele (e) which does not.

b ***What genotypes could be possessed by:***
1 the strongly cyanogenic plants
2 the weakly cyanogenic plants
3 the acyanogenic plants?

c ***Explain the results of part B of this investigation in terms of the allelic pairs G/g and E/e.***

d ***Methylbenzene (toluene) is a good solvent for lipids. It is effective in promoting the release of HCN from the strongly cyanogenic plants. Put forward a hypothesis to connect these two facts.***

INVESTIGATION
19C Gene expression in round and wrinkled peas (*Pisum sativum*)

(*Study guide* 19.1 'Mutant complementation in *Neurospora*'–see 'Mutations and metabolic pathways; and 19.9 'Epistasis'.)

If you have performed investigations on the inheritance of round and wrinkled seed characters in peas (see *figure 23*), you will have noticed that the two types of seed can occur in the same pod; so the difference is not merely the result of varying conditions in the environment during ripening. The two characters may be used conveniently for identifying the effect of two alleles of a gene for seed shape, where round (R) is dominant to wrinkled (r). The inheritance of round and wrinkled peas was first demonstrated by Mendel. You can go a stage further and see whether these superficial differences in the appearance of the peas may not be the outcome of some more fundamental kind of structural, physiological, or biochemical variation.

If there is time, you should undertake each of the following investigations but if not, complete one or more and pool your results with members of the class who have carried out the others between them. You should begin the third investigation ('Physiological differences between the two kinds of pea seeds') at the start of the practical session as it takes at least an hour.

Figure 23
Round and wrinkled peas.
Photograph, Duncan Fraser.

Differences in uptake of water between round and wrinkled pea seeds

Procedure A

1 Weigh out roughly equal quantities of round and wrinkled pea seeds.
2 Put each lot of seeds to soak separately in a good depth of water for between 24 and 36 hours.
3 Drain off the water, mop off any excess surface moisture with paper towel or blotting-paper, and weigh the seeds again.
4 Record the percentage change in mass in each case.

Differences in the starch grains of round and wrinkled pea seeds

Procedure B

1 Take a soaked pea and cut it in half.
2 Scrape the cut surface so that a very few fine scrapings fall into a drop of water on a microscope slide.
3 Stir the contents of the drop thoroughly to disperse the particles. Add a coverslip. Place a small drop of iodine in KI solution at the edge of the coverslip so that iodine can diffuse under it.
4 Examine the preparation under low and then high power, paying particular attention to the region where starch grains are just tinted by the diffusing iodine solution.
5 Some starch grains may be inside large cells. Most will have become free. Use an eyepiece graticule to measure 20 chosen grains at random. (You must decide on a procedure for randomly selecting grains.)
6 Repeat **1–5** with the other kind of pea.
7 Make large drawings to show the difference in structure of the starch grains from the two types of pea.

Physiological differences between the two kinds of pea seeds

The construction of starch grains in the developing seed requires the activity of enzymes to convert sugars to starch.

The breakdown of the starch grains during germination requires the action of enzymes to hydrolyse the starch. It might be that the activity of the enzymes is in some way linked with the round or wrinkled appearance of the seeds. To investigate this we can extract phosphorylases or other starch-forming enzymes from the two kinds of pea seed and study their action on glucose-1-phosphate. Such enzymes can act on glucose only after it has been combined with phosphate, hence the need to use glucose-1-phosphate rather than glucose alone.

Extraction of enzyme and its action on glucose-1-phosphate

Procedure C

1 Weigh out 2.5 g of dried, round peas.

2 Grind the peas in an electric grinder until they are a fine powder.

3 Add the pea powder to 10 cm^3 distilled water in a beaker and stir thoroughly. (If the mixture is very thick you may need to add more water.)

4 Centrifuge to obtain a clear solution, free from sediment. This may require 5 minutes at 2500 g. Test the supernatant with iodine solution to ensure that it is free from starch.

5 Carefully pour the supernatant into a labelled container.

6 Repeat steps **1** to **5** with 2.5 g of wrinkled peas.

7 You will need a Petri dish containing agar impregnated with glucose-1-phosphate. Draw a line across the bottom of the dish to divide it into two halves and mark one side R (round) and the other side W (wrinkled).

8 Using separate, clean pipettes for each extract, place four well spaced drops of the extract from round seeds on the side marked R and four from wrinkled seed extract on the side marked W.

9 After leaving the dish for 15–20 minutes in an incubator at 30–35 °C, add a drop of iodine solution to one of the drops on each side of the dish. Leave the two drops for a minute and then soak up the surplus liquid with absorbent paper. Look for any evidence of starch beneath the surface of the agar. Record your observations.

10 Repeat step **9** three more times, at 10-minute intervals, using a fresh pair of enzyme extract drops each time.

Questions

a ***Suggest how the difference in water uptake by the two seed types may be correlated with the appearance of the dry seed.***

b ***In what ways are the starch grains of the two seeds different? How could the differences help explain the differences between the appearances of the round and the wrinkled seeds?***

c ***Would you be justified, on the evidence available to you, in saying that the differences in the structure of the starch grains and in the overall appearance of the seeds are due to differences in enzyme activity?***

d ***What evidence would you need in order to suggest that the three phenotypic differences between round and wrinkled pea seeds were the result of the action of two different alleles of a single gene rather than a chance combination of independent traits?***

CHAPTER 20 POPULATION GENETICS AND SELECTION

INVESTIGATION

20A Models of a gene pool

(*Study guide* 20.2 'The Hardy–Weinberg model'.)

Organisms vary enormously in the number of offspring each is capable of producing, the rate of production of the offspring, the number of times an individual will mate, the number of different mates it will accept, the population from which mates are selected, and whether self-fertilization commonly occurs. All such variables influence the way in which the alleles at each locus in the species are distributed among succeeding generations and the variables are collectively called its breeding system (see *Study guide II*, 20.3, 'Inbreeding and outbreeding').

Consider an imaginary breeding system where all the individuals in the population, which has a fixed size, produce gametes simultaneously and deliver them into a small lake where they fuse in pairs, quite at random, by a process of external fertilization. A fixed number of zygotes, an entirely random selection of the vast number produced, develops into adults which reproduce by the same system at regular intervals. No organism has such a breeding system—it is approached most closely by certain marine organisms with external fertilization—but we can use it as a model of more complex real breeding systems and investigate the way a pair of alleles, A/a, is distributed among each generation of the imaginary organism. This may teach us something of the way real alleles are distributed in populations of real organisms. The model can be made into a physical model using objects to represent the alleles or into a computer model in which electronic signals represent the alleles (see *figure 24*).

Procedure

1 Let a bead (pea, marble, or any object) of one colour (for example, red) represent the dominant allele A and a bead of another colour (for example, blue) represent the recessive allele a. Single beads can also represent the haploid gametes containing the alleles.

2 Let a pair of beads (they can be physically joined if poppet beads are available) represent a diploid individual. This might be 2 reds (AA) or 1 red and 1 blue (Aa) or 2 blue (aa).

3 Label four containers as follows:

1 Pool of gametes;
2 Red homozygotes, AA;
3 Heterozygotes, Aa;
4 Blue homozygotes, aa.

Figure 24
Scoring beads and storing results on computer.
Photograph, Duncan Fraser.

4 Place 50 AA individuals, that is, 100 red beads, in the AA container and 50 aa individuals, that is, 100 blue beads, in the aa container. This is our founding population and it contains no heterozygotes.
5 Decide on a number of gametes an individual will produce at a mating, for example, 8. (It is much easier to calculate if an even number is chosen.)
6 Place the gametes in the container labelled 'pool of gametes'. There should be 50×8 red beads and similarly 400 blue beads.
7 Mix up the beads in the pool of gametes very thoroughly and then, without looking, take out 100 pairs, representing the 100 surviving zygotes.

8 As you remove each pair of beads record the genotype on a tally chart (see table 3) and put it in one of the containers you have prepared. (Make sure these are clear of the 'previous generation' first.)

Generation number	*2 red beads* (genotype AA)	*1 red and 1 blue bead* (genotype Aa)	*2 blue beads* (genotype aa)
1	卌 II	卌 卌 IIII	IIII

Table 3

9 After producing a total population of a hundred individuals examine the tally chart you have produced. Calculate the ratio of the genotypes in the population.

Questions

a ***Why would you expect the pairs of beads to be in the ratio of 1 red, red (AA): 2 red, blue (Aa): 1 blue, blue (aa)?***

b ***Why would you be surprised if the actual result was exactly 1:2:1?***

c ***Use the χ^2 test to see if your actual result departs significantly from the 1:2:1 ratio expected.***

10 Return to your bead model. Allow each individual to produce 8 gametes in the same way as the parents did. Calculation of the number of gametes produced by each group of homozygotes is very straightforward. If you understand meiosis you should also realize that the heterozygotes must produce equal numbers of each type of gamete, for instance, 4 red beads and 4 blue beads each.

11 Put the appropriate number of gametes of the two colours in the pool of gametes (which must be cleared of gametes from the previous generation.)

12 Mix thoroughly and withdraw 100 zygotes, recording them in the tally chart under generation 2 (see step **8** and table 3).

Question

d ***Does the ratio of diploid genotypes in generation 2 differ significantly from a ratio of 1 red, red (AA): 2 red, blue (Aa): 1 blue, blue (aa)?***

13 Using the numbers of individuals in generation 2, repeat steps **10–12** to obtain a third generation and record it in the tally.

14 Record your results for the three generations on a composite table of results for the whole class.

Questions

e ***Considering the results as a whole, in how many cases did a result depart significantly from a 1:2:1 ratio?***

f ***Does a significant departure mean that bias has been introduced (for example, by failing to mix the beads properly, or by miscounting) or would some significant departures occur by chance?***

g ***In the model, we have used a fairly large population (100 individuals and 800 gametes). Would the ratios produced have been very different if we had varied the total size of the population between certain limits, for instance, 50 and 200, as it moved from one generation to the next?***

h ***Provided that homozygotes always produced gametes of one colour and heterozygotes produced equal numbers of the two colours, would the number of gametes produced by each individual in the model have affected the ratio?***

Having answered the above questions, you should now understand the basic features of a gene pool or gamete pool model and can proceed to develop it using one of the several computer programs available. The program will allow you to select the size of population you wish to imagine and the number of generations for which you wish it to breed. You will perhaps already have decided that the smaller the population the greater the chance that the expected 1:2:1 ratio of genotypes will be disturbed. Small populations tend to 'drift' away from an expected genotype frequency and geneticists call this effect, which also happens in small laboratory populations of experimental organisms, genetic drift.

In the procedure above, you started with a frequency of 0.5 (that is, 50%) red beads (allele A) and 0.5 (50%) blue beads (allele a). Your computer program should allow you to select other allele frequencies, for instance 0.99 A and 0.01 a (as might happen if the allele a had just been introduced into a population). Exactly the same principles apply. The basic rules of the model are that each individual produces the same number of gametes, that the gametes successful in fusion are selected randomly, and that heterozygotes produce equal numbers of the two possible kinds of gametes.

The computer program may allow you to introduce a bias against one of the two types of allele. This is called selection. Individuals of one genotype, for example, AA, might be imagined to be less successful in reproduction than the other genotypes; for example, AA individuals might produce nine successful gametes for every ten successful ones produced by genotypes Aa and aa. It is fairly easy but very tedious to calculate how many beads of each colour to put into the gamete pool container if one makes this type of assumption. The computer will do it quickly and calculate what the expected allele frequency would be if the effect continued to happen for, say, ten generations. Clearly the frequency of allele A would decline and that of allele a would rise. Could the effects of chance, that is, genetic drift, in a very small total population, overcome a selection of 10% against allele A? A good computer program will allow you to explore such questions. You could do it with beads if you had the time and patience.

INVESTIGATION
20B Cyanogenesis and selection in *Trifolium repens* (white clover)

(*Study guide* 20.5 'Polymorphism'.)

Introduction

In practical investigation 19B, 'The biochemistry of cyanogenesis in *Trifolium repens*', you will have discovered that four different phenotypes of white clover can be identified and you will probably have discussed the possibility that the production of poisonous cyanides might deter animals from grazing on the clover or that the leaves themselves might be harmed by their own poison.

Cyanogenesis in clover is a good example of a genetic polymorphism. The idea of a polymorphism is discussed in *Study guide II*, section 20.5, 'Polymorphism'. Other polymorphisms exist in clover, as you may have discovered if you attempted investigation 17A, 'Describing variation'.

Procedure A The action of predatory herbivores

1 Identify two large clumps of clover one of which is definitely acyanogenic and the other strongly cyanogenic. The clumps should have arisen by vegetative reproduction so that all the leaves are of the same genotype. Several replicate picrate tests should have been carried out to be sure of the phenotype of each clump.
2 Detach stolons from each clump and push their bases into moist potting compost in small plant containers. Push the compost down firmly round each cutting with your fingers and water if necessary. Label the containers clearly, for instance + for strongly cyanogenic and − for acyanogenic.
3 Cover each cutting with an inverted jam jar, a polythene bag, or

other transparent cover. Leave it in a bright but cool place for about a week to become well rooted.

4 With scissors, trim off the older dead and dying or damaged leaves from each cutting and sort the plants into pairs of similar size, one cyanogenic and the other acyanogenic. If necessary, cut off some leaves from a large plant so that it matches a smaller one.

5 Count the number of leaflets (not leaves) on each plant in each pot. Make quite sure each leaflet is undamaged. Write the number on a record sheet and in felt pen on the side of the pot.

6 *Either* put all the paired pots in a large plastic bag, arranging them in a Latin square. Add one slug or snail per pot, blow the bag up, and tie the top firmly.
Or put one pot of each kind, choosing well matched plants, into a plastic box. Put a slug or snail in each pot. Put the lid on the box.

7 After about 24 hours, remove the pots and count the number of intact and the number of nibbled leaflets. Make a separate note of any leaf which has had its petiole nibbled through. (The petioles do not contain any cyanogenic glucosides and cannot produce cyanide.) See also *figure 25*.

8 Combine your results with other groups and record them in a table similar to table 4.

9 Carry out a chi-squared test to see if the molluscs used have significantly preferred one type of leaf to another.

Questions

a ***Is the degree of cyanogenesis the only way in which the leaves presented for choice by the molluscs varied?***

b ***Would the results of the experiment be more or less meaningful if the molluscs were unusually hungry at the start?***

c ***If other food plants had also been available, as they almost certainly would have been in a natural situation, how might this have affected the result?***

d ***What further investigations of predation of the different types of clover might be worth while?***

Procedure B The action of frost on Trifolium repens

1 Carefully take leaves with stalks from cyanogenic and from acyanogenic plants. Put several promptly into separate sets of very small specimen tubes or test-tubes containing a little water at the bottom for the cut bases of the petioles.

2 Keep the sets of acyanogenic leaves separate from the sets of

Date ....

Species of mollusc (if known) ....

Number of pairs of plants tested ....

	Cyanogenic plants				*Acyanogenic plants*			
	Nibbled leaflets	Intact leaflets	Total number of leaflets available	Petioles bitten through	Nibbled leaflets	Intact leaflets	Total number of leaflets available	Petioles bitten through
Observed number								
Number expected if chance alone determined choice								

Table 4

Figure 25
Snails feeding on clover. The rows marked + contain only cyanogenic plants. The other rows have only acyanogenic plants.
Photograph, Professor A. D. Bradshaw, Department of Botany, University of Liverpool.

cyanogenic leaves. You will need at least two lots of each sort of leaf. Enclose each lot of leaves in a slightly larger specimen tube so that only a small amount of air surrounds them.

3 With forceps, put a picrate paper with the leaves and take care to ensure that it cannot become wet during thawing (see *figure 26*). It is

important to have a large volume of leaves compared with that of the atmosphere into which they can release hydrogen cyanide. This enables small quantities of the gas to be detected.

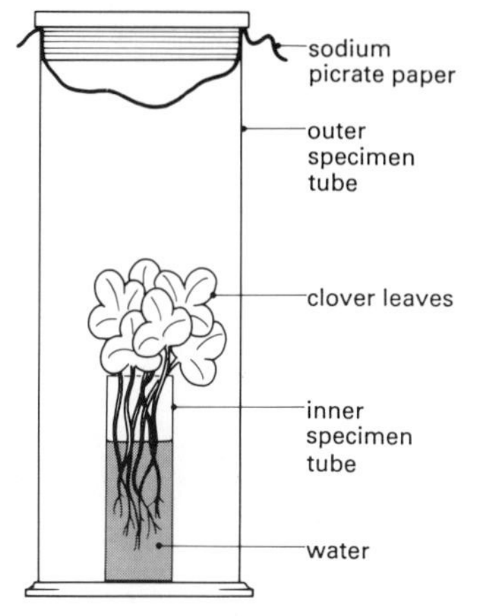

Figure 26
Possible arrangement of clover leaves for treatment with frost.

4 Put one lot of each type of leaf in the main compartment of a domestic refrigerator and the other lot in the freezing compartment or into a deep freeze. Record the temperature of these two locations. The leaves should be left for between one and three hours.

5 Remove the tubes of leaves, check that the picrate paper cannot become wet during thawing, and leave the tubes in a dark place at room temperature for twenty-four hours.

6 Examine the tubes and record the results. Look carefully at the leaves themselves and note their appearance, as well as that of the picrate paper.

Questions

e ***What do the results suggest about the effect of low temperatures on the leaves of the two sorts of clover plant?***

f ***Look carefully at* figure 27. *To what extent do your results suggest a reason for the data represented by the figure?***

g ***Imagine that there is a mixed population of cyanogenic and acyanogenic clovers growing at around 700 m in a mountainous region. Explain the effects that predation and climatic conditions could have upon the evolution of the clover population in that region.***

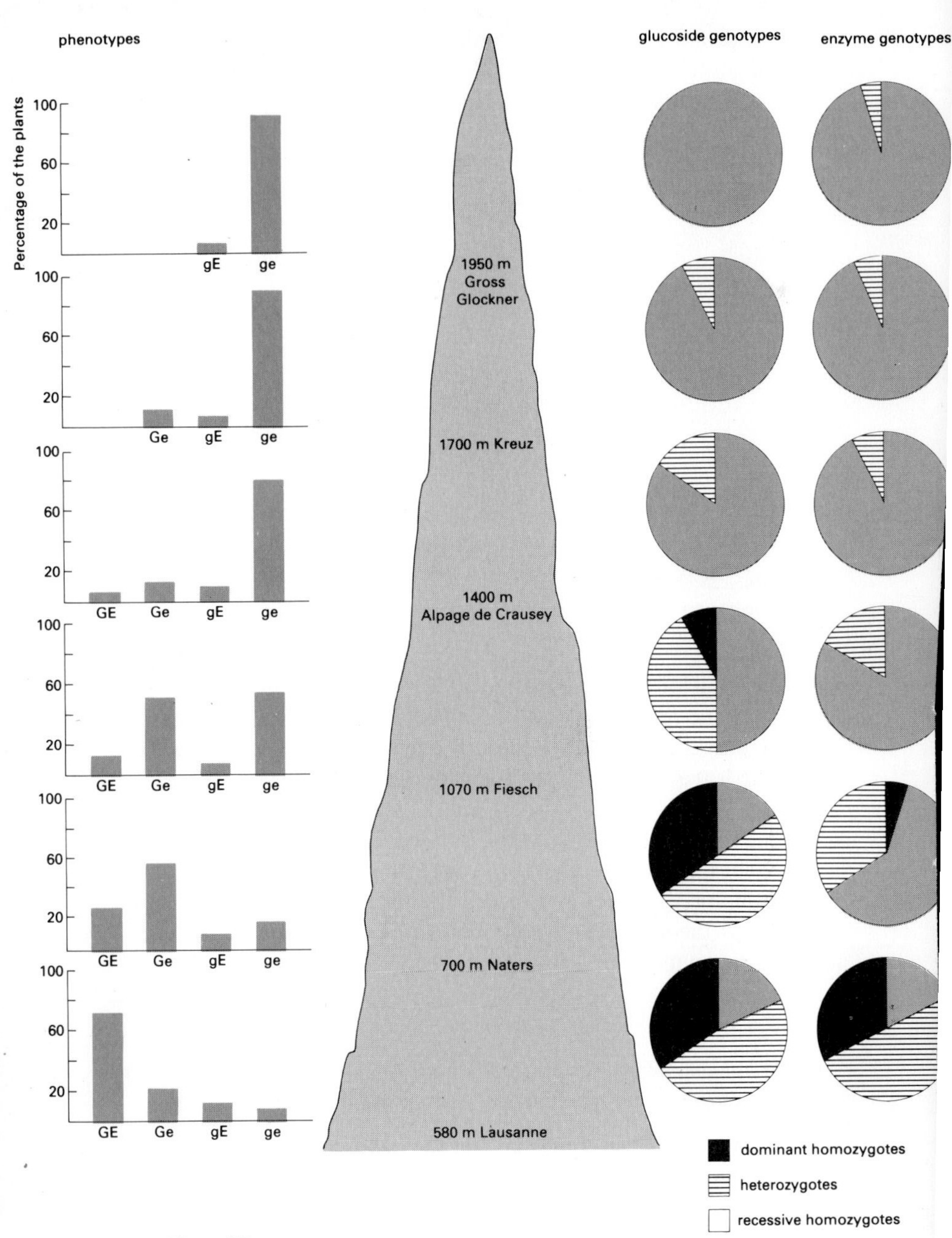

Figure 27

Phenotype and genotype frequencies in wild populations of *Trifolium repens* from different altitudes.

From Daday, H. 'Gene frequencies in wild populations of Trifolium repens', *Paper No. II 'Distribution by altitude'*, Heredity, **8**, *1954, pp. 377-84.*